Building a Data Culture

The Usage and Flow Data Culture Model

Gary W. Griffin
David Holcomb

Apress®

Building a Data Culture: The Usage and Flow Data Culture Model

Gary W. Griffin
Jacksonville, FL, USA

David Holcomb
Berryville, VA, USA

ISBN-13 (pbk): 978-1-4842-9965-4
https://doi.org/10.1007/978-1-4842-9966-1

ISBN-13 (electronic): 978-1-4842-9966-1

Managing Director, Apress Media LLC: Welmoed Spahr
Acquisitions Editor: Shivangi Ramachandran
Development Editor: James Markham
Editorial Project Manager: Jessica Vakili

Cover designed by eStudioCalamar

Distributed to the book trade worldwide by Springer Science+Business Media New York, 1 New York Plaza, Suite 4600, New York, NY 10004-1562, USA. Phone 1-800-SPRINGER, fax (201) 348-4505, e-mail orders-ny@ springer-sbm.com, or visit www.springeronline.com. Apress Media, LLC is a California LLC and the sole member (owner) is Springer Science + Business Media Finance Inc (SSBM Finance Inc). SSBM Finance Inc is a **Delaware** corporation.

For information on translations, please e-mail booktranslations@springernature.com; for reprint, paperback, or audio rights, please e-mail bookpermissions@springernature.com.

Apress titles may be purchased in bulk for academic, corporate, or promotional use. eBook versions and licenses are also available for most titles. For more information, reference our Print and eBook Bulk Sales web page at http://www.apress.com/bulk-sales.

Any source code or other supplementary material referenced by the author in this book is available to readers on GitHub. For more detailed information, please visit https://www.apress.com/gp/services/source-code.

Paper in this product is recyclable

This book is dedicated to our wives, Tammy Griffin and Rhonda Holcomb. You put up with a lot over the years. We are truly blessed to have you by our side. Thank you for all of your support, even when it wasn't easy.

Table of Contents

About the Authors..xi

Introduction xiii

Foreword xvii

About the Foreword Author xix

Chapter 1: Data Culture 1

What Is Culture?.. 3

Introduction to Organizational Culture 4

Culture and Culture Change 7

A Data Culture Perspective 10

Summary... 15

Chapter 2: The Current State 17

Usage and Flow Data Culture Model (UFDCM) 18

Organizational Structures 26

Components of a Mature Data Culture... 37

Charting Your Journey.. 42

Summary... 49

Chapter 3: Organizational Vision and Data Strategy 51

Why Do We Need a Data Strategy? 52

Data Strategy and Organizational Strategy... 54

Data Strategy and Data Culture Interaction 56

Overcoming Challenges in Data Culture Transformation................................ 57

Data Strategy and Talent Development.. 59

A Survey of Data Culture Within Each Quadrant... 60

Data Strategy in a Preservationist Culture 61

Data Strategy in a Protectionist Culture 62

Data Strategy in a Traditional Culture 64

Data Strategy in a Progressive Culture 67

Highly Regulated Industry Data Strategy (A Protectionist Example) 69

Summary 71

Chapter 4: Leadership and Change Management **73**

Organizational Structures and the Data Governance Office 74

The Role of Leadership in Driving Change and Data Transformation 79

Building a Coalition of Champions and Advocates 80

Strategies for Effective Change Management 82

Fostering a Supportive Environment for a Data Culture 83

Summary 85

Chapter 5: Data Governance and Infrastructure **87**

Data Culture and Data Governance 88

Data Governance: Librarian Initiatives 91

Modern Data Catalog (Cornerstone of Librarian Services) 92

Data Governance: Compliance/Security 94

Data Governance: Quality 96

Data Governance: Process Initiatives 97

Data Governance: Communication 98

Data Governance: Training 99

Data Governance in Each Culture Type 101

Data Governance in a Progressive Culture 101

Data Governance in a Preservationist Culture 102

Data Governance in a Traditional Culture 104

Data Governance in a Protectionist Culture 105

Summary 107

Chapter 6: Data Literacy and Skills Development **109**

Promoting Data Literacy in a Data Culture 110

Designing and Implementing Data Training Programs 111

Continuous Learning and Data Skills 113

Data Literacy in a Traditional Culture 115

Data Literacy in a Protectionist Culture 116

Data Literacy in a Preservationist Culture 118

Data Literacy in a Progressive Culture 119

Triggers for Data Literacy 120

Summary 121

Chapter 7: Embedding Data into Decision-Making **123**

Data Analytics and the Data Culture Model 125

Leveraging Data Analytics for Insights 125

Balancing Data Flow for Effective Analytics 128

Data Privacy and Security Within the Data Culture Model 130

Establishing Robust Data Privacy Measures 131

Impact of Data Privacy and Security on Data Usage 132

Building Trust and Democratization Through Security 133

Analytics, Privacy, and Security in a Traditional Data Culture 134

Analytics, Privacy, and Security in a Protectionist Data Culture 136

Analytics, Privacy, and Security in a Preservationist Data Culture 138

Analytics, Privacy, and Security in a Progressive Data Culture 140

Summary 142

Chapter 8: Nurturing Communication and Collaboration **145**

Focus, Collaboration, and Communication Interplay 146

A Mature Data Culture Enhances Communication and Collaboration 148

Communication, Collaboration, and Breaking Down Silos 152

Communication and Collaboration Between Data Professionals and Business Stakeholders.. 154

Communication and Collaboration in the UFDCM 156

Communication and Collaboration in a Progressive Culture..................... 156

Communication and Collaboration in a Traditional Culture 159

Communication and Collaboration in a Preservationist Culture................ 162

Communication and Collaboration in a Protectionist Culture................... 165

Summary.. 168

Chapter 9: Measuring Success and Sustaining the Data Culture 171

Establishing Metrics and Indicators... 173

Aligning Metrics with Organizational Objectives..................................... 176

Measuring Progress and Success... 178

Evaluating Data-Driven Decision-Making 179

Assessing Data Culture Adoption 180

Continuously Refining the Data Culture Strategy.................................... 182

Adapting to Organizational Changes.. 184

Strategies for Sustaining the Data Culture 185

Summary.. 186

Chapter 10: Case Studies and Lessons Learned 189

Industries and the UFDCM 190

Progressive Culture... 191

Traditional Culture... 192

Preservationist Culture.. 193

Protectionist Culture 194

Case Studies and Experiences... 196

Case Study #1 197

Case Study #2 199

Case Study #3 202

Case Study #4 206

Summary.. 208

Bibliography **211**

Index.. **219**

About the Authors

 **Gary W. Griffin** has worked for over 30 years at all levels of public and private organizations including the Centers for Disease Control, Georgia Department of Public Health, Georgia Department of Education, Atlanta Public Schools, Gwinnett County Public Schools, FedEx, Lockheed Martin, Woolworths, ConocoPhillips, Ahold USA, IBM, and Hills Department Stores, to name a few. His duties have ranged from data collection and reporting to strategic planning and leadership. His roles have changed dramatically over the years, but most recently he spent more than 20 years working as Director of Data Quality and Research Accountability at Atlanta Public Schools, a managing consultant at IBM, and as the founder and CEO of TASADA Solutions. These roles had one thing in common; he was engaged in delivering enterprise solutions that focused on data, analytics, and strategy.

Dr. Griffin is a highly skilled leader with a primary focus and expertise in data strategy, data governance, and advanced analytics that spans more than 30 years. In fact, he worked on a team while at IBM to help build one of the first data warehouses in a state department of education in South Carolina in 1998. He is the inventor of Datalink 1000, a Master Data Management and Data Quality tool. He also developed the data strategy for one of the first web-based assessment systems for K-12 education.

Dr. Griffin has a Master of Science from Auburn University and a Doctor of Philosophy in Sociology from Georgia State University. He is a highly skilled applied social scientist and a data culture change agent. He is a noted author, speaker, educator, entrepreneur, innovator, and visionary leader in the field of data, analytics, and strategy.

 **David Holcomb's** career includes leadership roles in business operations, marketing, sales, product development, and information technology with revenue and cost center responsibilities in leading brands including Verizon, Capital One, DHISCO, Western Union, and the University of Chicago Medical Center. Through this diversity of experience, he has developed a deep appreciation for the unique perspectives and uses of data across an organization. These perspectives have allowed him to lead corporate transformations including cloud migrations, process and organizational redesign, startup formation, and real-time prescriptive analytics development. Dr. Holcomb has also led organizational, data, and process integrations in multiple global mergers and acquisitions.

His academic experience includes teaching, curriculum development, doctoral committee membership, and speaking. He has taught online and on-ground at the undergraduate and graduate levels at multiple institutions in information management, business, supply chain, organizational behavior, and technology risk management. In curriculum development, Dr. Holcomb redesigned supply chain and data warehousing coursework and developed the capstone course for a bachelor's degree in project management. He has also performed as a doctoral dissertation committee member, specifically as the subject matter expert for leadership theory.

Dr. Holcomb holds a Bachelor of Arts degree from Western Illinois University, a Master of Science in Management Information Systems from Nova Southeastern University, a Doctor of Philosophy in Business Administration from Touro University International, and a Doctor of Philosophy in Information Technology from the University of the Cumberlands. He is a speaker, theorist, and writer in data warehousing, business intelligence, organizational behavior and culture, human behavior, and cyber security.

Introduction

Data culture is a broad topic to cover in a single book. This book explores the intricacies of building a thriving data culture within organizations. It delves into the fundamental principles, strategies, and best practices that enable organizations to harness the power of data to drive innovation, make informed decisions, and deliver value to customers.

Chapter 1 begins with Peter Drucker's famed saying, "Culture eats strategy for breakfast." We continue with an introduction to culture by providing a few definitions from different sources. It begins with an explanation of organizational culture and lands on data culture; what it is and how it relates to organizational culture. A brief introduction to the Usage and Flow Data Culture Model is provided to readers to set the tone and direction for the rest of the book.

In Chapter 2, the critical task of assessing the current state of an organization's data landscape is covered. We explore how organizations can evaluate their existing data infrastructure, processes, and practices to identify gaps and challenges that hinder the development of a data culture. Furthermore, we delve into understanding the organization's readiness for change and the potential barriers that need to be overcome to embark on a successful data transformation journey.

Chapter 3 explores the elements of organizational vision and data strategy in building a data culture. A compelling vision aligning with the organization's overall goals and objectives is paramount to success. We investigate the importance of developing a data strategy that outlines the steps and initiatives required to drive the desired data culture. Additionally, we emphasize the significance of effectively communicating the vision and strategy to gain buy-in and support from stakeholders at all levels.

The pivotal role of leadership in driving culture change and data transformation takes center stage in Chapter 4. We look at the influence of leadership in shaping and nurturing a data culture. We explore strategies for building a coalition of champions and advocates to drive the data culture initiative forward. Furthermore, we provide insights into effective change management practices, including addressing resistance and fostering a supportive environment for embracing a data-driven mindset.

INTRODUCTION

Chapter 5 dives into the critical aspects of data governance and infrastructure. A robust data governance framework ensures the confidentiality, integrity, and accessibility of data. Building or enhancing data infrastructure to align with the organization's data culture goals is also explored. Additionally, we discuss the effective utilization of technology and tools to facilitate efficient data management and accessibility.

Data literacy and skills development are the focus of Chapter 6. We emphasize the importance of promoting data literacy among employees at all levels of the organization. We discuss the design and implementation of data training programs and resources that enable individuals to develop the necessary skills to work with data effectively. Moreover, we highlight the significance of fostering a continuous learning culture to enhance data skills and capabilities.

In Chapter 7, we examine the integration of data-driven decision-making into organizational processes and workflows. We describe how organizations embed data into their decision-making processes to drive better outcomes and achieve strategic objectives. Additionally, we discuss how security and privacy interact with analytics within the data culture. Analytics is clearly tied to Data Usage, while data security and privacy are directly linked to Data Flow.

In Chapter 8, we turn our attention to the topics of collaboration and communication to encourage cross-functional interactions and break down silos that can be a severe bottleneck to both the usage and flow of data within a data culture. Establishing communication channels and platforms for sharing data insights and promoting a data-sharing culture that fosters collaboration between data professionals and business stakeholders can't be underemphasized for a mature data culture.

It's necessary to measure success to sustain a thriving data culture. Establishing metrics and indicators to measure the progress and success of the data culture initiative provides a continuous feedback loop. Continuously refining and adapting the data culture strategy based on feedback and results and a discussion of strategies for sustaining the data culture in the long term are the focus of Chapter 9.

In the final chapter, we focus on real-world experiences from the field that illustrate different data cultures encountered over the years. Here, we present real-world case studies of successful and not-so-successful data culture transformations. We highlight challenges faced and lessons learned throughout the process and provide readers with valuable insights and actionable takeaways.

This book is primarily written for professionals and leaders across various industries who are interested in building a data culture within their organizations. The typical reader may have a background in data management, analytics, business intelligence, or technology, but the book is designed to be accessible to a wide range of readers with varying levels of expertise. Overall, the book caters to a diverse audience interested in building a data culture, including both technical and nontechnical professionals. It provides practical guidance, real-world examples, and a comprehensive framework to help readers develop the necessary skills and knowledge to foster a mature data culture within their organizations.

Foreword

During the beginning of the COVID pandemic in 2020, David Holcomb was working on his second PhD. Suddenly, the world was stopped, and working from home became a reality in weeks. Using his experience with social exchange theory, David wrote his thesis on using technology to replace the in-office experience during the lockdown, how remote work would influence corporate culture change, and the impacts on work relationships. When David started to collect data for his research, he met (virtually) Gary Griffin in a social media forum. Like David, Gary had decades of being a practitioner in operations, data management, and analytics and a PhD in Sociology. The authors bring over 70 years of practical experience as executives and consultants for different industries and markets. The more conversations Gary and David had, the more they could see the similarities of their ideas, how their line of thinking complemented each other, and, most importantly, what they believed most enterprises were missing when looking at data to make high-impact decisions. Gary and David continued having regular discussions about these topics, even after David finalized his dissertation. And that's when the idea of crafting this book started to flourish.

Why did Gary and David decide to spend time in their busy lives writing this book about data culture? Financially, they did not need to embark on such an extreme journey. For everyone who wrote or tried to write a book, it means long days, no free weekends, and less time with family and friends. Gary says it is about his legacy, to leave a blueprint to the world on how to succeed with data and analytics. For him, it is his contribution to the society. On the other hand, David loves to teach and becomes excited when he sees his knowledge multiply. For him, the sooner the better. Both agree that if they had a book like this one when they started their careers, they would have a blueprint to follow, and they would have made fewer mistakes.

Why do we need another book on culture? Haven't enough materials been published on culture, people, and influence so far? Before you answer, please let me provide you with another data point. Like a language, culture is not static but changes through time and context. A few items that would be considered a norm today in the United States, like women voting in government elections, were not a norm until the 1920s (in some countries, even later than here). A book exploring the concept of building and managing data culture like this one is considered a seminal work in data and analytics. The book drills down how the data culture must be influenced by organizational culture, how to craft a data strategy considering the cultural component, and how to consider data governance, data quality, data privacy, and data literacy as pillars to support Data Usage to decision making. After all, people are the weakest link in the chain of **Process-Technology-Data-People**.

Culture is like a beast; it will eat you alive if you do not apply the correct principles to learn and behave appropriately. In 2015, when I was at Gartner doing my research, I wrote a paper that, in the years to come, Chief Data Officers (CDOs) would become mainstream in corporations and that 50% of them would fail, listing several reasons why. Unfortunately, I was correct in my prediction and analysis of the CDOs. In the years after I wrote my research, I had the opportunity to work with many companies worldwide. I learned that one of the main reasons data and analytics leaders failed was because they were trying to bring to their companies changes the C-Level executives needed more time to be ready to understand, accept, embrace, or support. It generated frustrations, leading the CDOs to pursue other career opportunities.

This book will resonate with each one of us who are being tasked to use data in business decisions, to create new products, to promote change, and to drive efficiency. Following the guidelines will allow you to be consistent in your approach when working with your team, peers, leaders, and stakeholders. Each chapter of the book has the potential to become a book on its own; so deep are the topics and ideas explored there. The beauty of what Gary and David did is a summarization of several interconnected and interdependent areas of data and analytics, tied together by the Usage and Flow Data Culture Model (UFDCM), a framework to help you understand how to look at the company you are currently inserted and the steps you need to take to drive your success. Like the authors, I wish I could have had a book like this one when I started my career as a Chief Data Officer.

Enjoy your reading, and good luck in your data and analytics journey.

Mario Faria

Professor at Carnegie Mellon University

Chief Data and Analytics Officer Program, Heinz College

About the Foreword Author

Mario Faria is a professor at the Heinz College at Carnegie Mellon University, Chief Data Officer, and a global advisor for companies looking to drive value by creating and executing data, analytics, and digital transformation projects and programs.

Mario works along several business lines to define the goals, priorities, and strategies for the following: growth opportunities, product offerings, markets to pursue, and competitors to look at. He has professional experience at Amazon, Gartner, Equifax, Accenture, and Microsoft, leading global teams in data, analytics, technology, CRM, supply chain, post-merger integration, and operations management.

Mario is considered one of the world's first chief data and analytics officers and the first one in Latin America, and he has been involved with the MIT CDOIQ Symposium and working with CDOs since 2011.

Mario has a BSc in Computer Science from Unicamp (Brazil), an MSc in Computer Science from the State University of New York at Albany, and an MBA in Marketing from the University of California at Santa Cruz.

CHAPTER 1

Data Culture

It has often been said, "Culture eats strategy for breakfast." Peter Drucker, a noted 20th-century management philosopher and consultant, has been credited with this infamous saying. He highlights the significance of organizational culture in shaping the success or failure of strategic initiatives. Drucker emphasized that no matter how well crafted a strategic plan may be, an organization's underlying culture determines its ability to execute that strategy effectively.[1]

In this context, "culture" refers to the shared values, beliefs, norms, and behaviors defining an organization and its operations. It encompasses the attitudes and behaviors of employees, the relationships between different departments or teams, and the overall working environment. Culture influences how people within an organization approach their work, make decisions, collaborate, and respond to challenges.

The most brilliant strategy will struggle to succeed if it is not aligned with the existing organizational culture or if the culture itself does not support the desired outcomes. In other words, if there is a clash between the strategic goals and the prevailing culture, the culture will significantly impact the organization's performance. For example, if an organization has a culture that discourages risk-taking and innovation, a strategy that relies on rapid experimentation and adaptation may face significant resistance. Similarly, if the culture promotes individualism and siloed thinking, a strategy to foster collaboration and cross-functional teamwork may struggle to gain traction.[2]

In his work, Drucker focused heavily on the information worker and using data in decision-making. Drucker was a proponent of data democratization and managers

---

[1] See Peter Drucker (1991). `https://drucker.institute/did-peter-drucker-say-that/`, Drucker never said "Culture eats strategy for breakfast." Drucker did say "Culture—no matter how defined—is singularly persistent."

[2] We take the position that culture is the most decisive factor in building a data culture. Failures, or limited successes, in endeavors with the desired outcome of building a data culture can be directly attributed to culture. That is to say, it is largely people and behavioral issues that determine success.

G. W. Griffin and D. Holcomb, *Building a Data Culture*, https://doi.org/10.1007/978-1-4842-9966-1_1

having access to data to increase the accuracy of decisions by reducing assumptions or using opinions. He advocated getting data to the front line by implementing key performance indicators (KPI), identifying trends, and gaining insights into customer behavior and market dynamics. With this data-sharing strategy, Drucker believed decision rights could be moved closer to the operational activities, making the organization more responsive and efficient.[3]

Expanding on Drucker's "culture eats strategy for breakfast" philosophy, a "data culture" must also align with the organizational culture and support the strategy. Recalling our previous example, if the culture discourages risk-taking and innovation, the data culture must support high accuracy to support these behaviors, or the organizational strategy will not reach its potential. This high level of accuracy may come at the cost of speed of access or breadth of data available. This alignment of data culture does not mean data is always available but that the data culture (usage and flow) is consistent with the organizational culture and strategy.

Drucker's assertion regarding culture and strategy implies that leaders should pay attention to the cultural aspects of their organization and, if necessary, shape or change the culture to create an environment conducive to successfully implementing strategic initiatives. With the evolution of advanced analytics, data storage, access methods, computing capacity, and tech savviness of the user community, leaders must define a data culture aligned with the organizational culture. Organizations with complementary organizational and data cultures will reduce friction in decision-making and be better positioned to realize strategic outcomes.

This chapter continues with definitions of culture, which will further be expanded to define organizational culture and ultimately data culture later in the chapter. Followed by this foundation, a treatment of organizational culture is presented. A data culture must support the organizational culture, so following the explanation of organizational culture, an introduction to the Usage and Flow Data Culture Model (UFDCM) offers a perspective of the competing values associated to data culture. This chapter acts as an introduction to the entire work with overviews of the key concepts for expanded treatment throughout the book.

---

[3] See Turriago-Hoyos, A., Thoene, U., and Arjoon, S. (2016)–"Drucker (1994, 1999, 2005) points out that this period of outstanding social and technical transformation evolved during a particular moment in the history of humanity that he labels "Knowledge Society." Nevertheless, Machlup (1962) documented empirically for the case of the United States that the knowledge society was created by a shift in the nature of work."

What Is Culture?

German poet, translator, and editor Hans Magnus Enzensberger said, "Culture is a little like dropping Alka-Seltzer into a glass of water; you don't see it, but somehow it does something." To begin the exploration of data culture, we must understand the idea of culture in general and organizational culture specifically. Additionally, the larger industry and national cultures will influence an organizational culture.

So then, what is culture? Culture has many definitions and different meanings. To see these differences, one need only review the American Heritage Dictionary, Dictionary. com, and Wikipedia page on culture. From these sources, the following passages were used to distill a list of elements of culture:

> "The arts, beliefs, customs, institutions, and other products of human work and thought considered as a unit, especially regarding a particular time or social group;
>
> These arts, beliefs, and other products considered with respect to a particular subject or mode of expression;
>
> The set of predominating attitudes and behavior that characterize a group or organization."[4]
>
> "The quality in a person or society that arises from a concern for what is regarded as excellent in arts, letters, manners, scholarly pursuits, etc.;
>
> That which is excellent in the arts, manners, etc.;
>
> A particular form or stage of civilization, as that of a certain nation or period: Greek culture;
>
> Development or improvement of the mind by education or training."[5]
>
> "Culture is an umbrella term which encompasses the social behavior, institutions, and norms found in human societies, as well as the knowledge, beliefs, arts, laws, customs, capabilities,

---

[4] American Heritage Dictionary, 2023
[5] Dictionary.com, 2023

and habits of the individuals in these groups. Culture is often originated from or attributed to a specific region or location."[6]

"A cultural norm codifies acceptable conduct in society; it serves as a guideline for behavior, dress, language, and demeanor in a situation, which serves as a template for expectations in a social group. Accepting only a monoculture in a social group can bear risks, just as a single species can wither in the face of environmental change, for lack of functional responses to the change. Thus, in military culture, valor is counted as typical behavior for an individual and duty, honor, and loyalty to the social group are counted as virtues or functional responses in the continuum of conflict. In the practice of religion, analogous attributes can be identified in a social group."[7]

These passages offer a great list of elements to consider when dealing with the concept of culture. These components include time, a specific group, and content for establishing norms. This time precept implies that cultures can and do change over time. For example, the culture of a small group will change as the group gets larger.

In addition to these passages and the elements noted, the working definition of "culture" for this book will integrate a quote from noted theorist Edgar Schein who said, "Culture is the way in which a group of people solve problems and reconcile dilemmas." For this book, the definition of culture is "the shared values, beliefs, attitudes, customs, and practices of a group of people to solve problems, reconcile dilemmas, and realize outcomes."

Introduction to Organizational Culture

With a working definition of culture, we can better understand organizational culture. Since "culture" refers to the shared values, beliefs, attitudes, customs, and practices, "organizational culture" refers to the shared values, beliefs, customs, and practices that characterize a specific organization. It is an organization's personality, representing how things are done, how people behave, and how decisions are made. Organizational culture is shaped by various factors, including the organization's history, mission, goals, leadership style, and the values and beliefs of its members.

---

[6] Edward Tyler, 1871

[7] Jackson, Y. Encyclopedia of Multicultural Psychology, p. 203

Cameron and Quinn from the University of Michigan proposed organizational culture models such as the "competing values" model (Figure 1-1).[8] Organizations will move from family owned (clan) or startups (adhocracy), where flexibility and taking risks are valued, to becoming more market focused (market) or repetitive (hierarchy), where stability and control are valued. This culture change requires different approaches to managing the organization to ensure performance. The adhocracy and market cultures tend to look externally to differentiate themselves, whereas the clan and hierarchy cultures focus inwardly on efficiency.

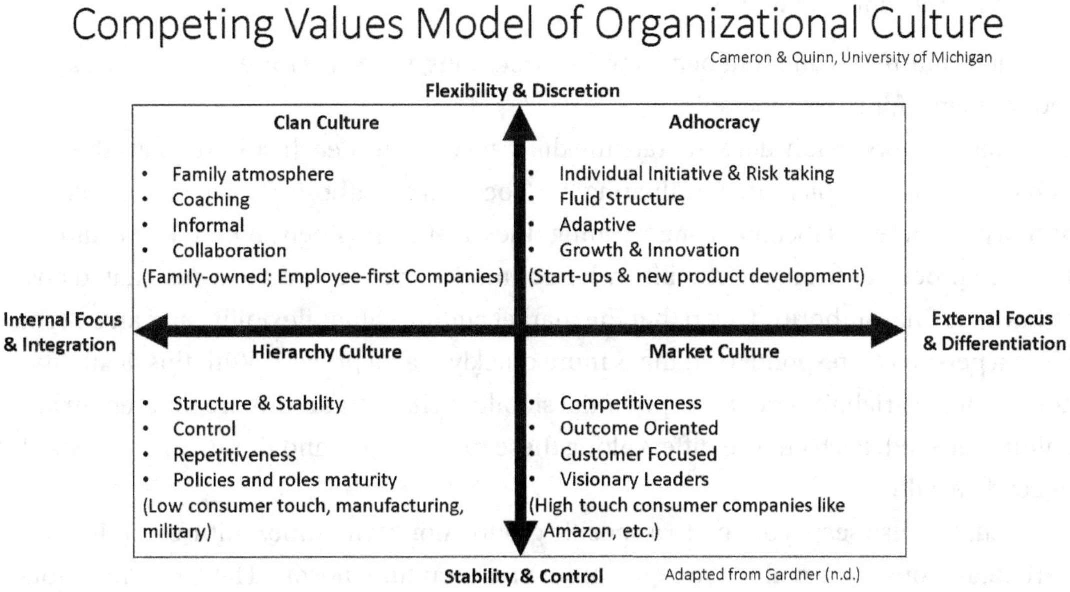

Competing Values Model of Organizational Culture

Cameron & Quinn, University of Michigan

Flexibility & Discretion

Clan Culture	Adhocracy
• Family atmosphere	• Individual Initiative & Risk taking
• Coaching	• Fluid Structure
• Informal	• Adaptive
• Collaboration	• Growth & Innovation
(Family-owned; Employee-first Companies)	(Start-ups & new product development)

Internal Focus & Integration ←→ **External Focus & Differentiation**

Hierarchy Culture	Market Culture
• Structure & Stability	• Competitiveness
• Control	• Outcome Oriented
• Repetitiveness	• Customer Focused
• Policies and roles maturity	• Visionary Leaders
(Low consumer touch, manufacturing, military)	(High touch consumer companies like Amazon, etc.)

Stability & Control   Adapted from Gardner (n.d.)

Figure 1-1. *Organizational Culture of Competing Values*

The transactional expectation and governing rules within each culture offer insight into the difference in the operation of each cultural quadrant. In a predecessor work to the competing values model, Quinn and McGrath described these transactional expectations and governing rules across nine vectors. These vectors are as follows:

1. Organizational Purpose

2. Criteria of Performance

---

[8] See Gardner (n.d.)

3. Location of Authority

4. Base of Power

5. Decision-Making

6. Leadership Style

7. Compliance

8. Evaluation of Members

9. Appropriate Motives

Each culture has different behavioral expectations, infrastructural placements, and requirements for its membership.

These vectors vividly demonstrate the differences in how each culture treats the behaviors. For example, when evaluating the "location of authority," each culture places authority in different locations or governing rules. Not surprisingly, hierarchical culture relies on processes to ensure stability, whereas market culture looks to a designated boss to have the final authority. Given that the market culture values flexibility and an external view, a person can respond to changes more quickly than a process. With this flexibility comes more variability and lower potential stability. These trade-offs are inherent in any culture selected. Each culture differs along these nine vectors, and these differences will affect data culture

A culture also seems bound to a specific group. Along with time, culture requires a particular group to be defined for agreement on the cultural norms. The group monitors the standards and takes corrective action if an actor violates them. Individuals new to a cultural group routinely violate standards and will be reprimanded or corrected. Cultural rules do not apply if a person is not part of the culture; for example, a citizen of France in France would not be subject to the cultural rules of China while in France. If a citizen of France were in China, the Chinese culture would govern interactions. Similarly, the culture at IBM may be quite different from the culture at Meta. Cultural groups can be defined in many ways, including geographic, demographics, religious, and others.[9]

The content used to establish norms of culture also has many facets. The list includes arts, beliefs, institutions, history, rituals, products of human work, behaviors, predominating attitudes, laws, and customs. The content of each of these facets can

---

[9] Fully discussed here: American Heritage Dictionary, 2023; Dictionary.com, 2023; Wikipedia, 2023

differ from organization to organization. As mentioned earlier, IBM and Meta could have distinctly different cultures based on their history, customs, or focus. The differences can be attributed to the legacy (history) and customs enacted before any current employees were part of the company.

The quadrants represented in Figure 1-1 are meant to be explanatory. These cultures exist to varying degrees across organizations corresponding to the degree of adoption of each vector. Models using these competing values create different cultural designations based on the point on each continuum. It should also be noted that "subcultures" or "micro-cultures" such as departmental, regional offices, or cliques can exhibit distinct points on the continuums and act independently of the espoused organizational culture. For example, the marketing department of a lumber company may have a different level of external focus than its manufacturing arm.

Culture and Culture Change

As organizations change, the organization's culture will need to change to fit the organization's maturity or the desired organizational state. As an organization changes from its founding to its current form, the culture must adapt to the changing demands. Many startups and family-owned businesses have had trouble transitioning to these later phases of the organizational life cycle. During the life cycle, the organization may reach a time of decline that requires a new direction or reinvention of the company. The organization may also reach plateaus requiring renewal to reinvigorate performance. Corporations and other organizations have a "life cycle" like any other organism.

One common model for the organizational life cycle is the startup-to-growth model, which outlines the different phases a company typically experiences. Let's discuss each phase in detail:

- Startup Phase: The startup phase is the initial stage of a company's existence. It begins with an idea or concept and is characterized by high uncertainty and risk. Startups typically focus on developing a minimum viable product (MVP) and validating their business model. The organization is small, with a flat structure, and the founders play multiple roles. The primary objective is to gain market traction and secure funding. In the case of the family-owned business, the goals may be to establish a reputation in the communal marketplace to make the business "minimally viable" to sustain the family.

- Growth Phase: If a startup successfully establishes a product-market fit and achieves sustainable revenue growth, it enters the growth phase. Similarly, if a family-owned business stakes a claim in a community, it may also move into the growth phase. This stage is marked by rapid expansion and increasing market presence. The company begins to hire more employees, open more outlets, build functional departments, and develop formal processes and systems. It requires effective leadership, strong management, and a focus on scaling operations while maintaining customer satisfaction.

- Maturity Phase: As the company continues to grow, it enters the maturity phase. The organization has achieved stability and a strong market position at this stage. It has established a recognizable brand and a loyal customer base. The focus shifts to optimizing operations, increasing efficiency, and sustaining profitability. The company may expand into new markets or product lines and establish strategic partnerships.

- Decline or Renewal Phase: In the decline phase, the organization experiences a decline in growth and faces challenges such as increased competition, changing market dynamics, or product obsolescence. If the company fails to adapt and renew itself, it may face a decline in market share and profitability. However, with strategic planning and innovation, organizations can rejuvenate themselves and enter a renewal phase. This phase involves reinventing the business, exploring new opportunities, and adapting to changing market conditions.

The organizational life cycle model provides a general framework, and the duration of each phase can vary widely depending on factors such as industry, market conditions, and management decisions. Additionally, not all organizations follow a linear progression through these phases. Some may experience rapid growth and maturity, while others may face challenges and regress. Likewise, an organization may experience a renewal phase multiple times to adapt to changes in the market, desire for new markets, or plateau in performance. Navigating an organization and its members through this life cycle requires deliberate and sustained leadership attention.

An organization's culture can significantly impact its performance as it shapes how employees interact with each other, approach their work, and respond to challenges and opportunities. A positive culture can foster collaboration, innovation, and employee engagement, while a negative culture can lead to low morale, high turnover, and poor performance. During a company's transition between phases, a positive culture is vital to preserve its place and progress with its maturation. Therefore, it is important for organizations to actively manage their culture and work to create a positive and supportive environment for their employees.

Implementing a culture change within an organization can be a complex and challenging process, but several steps can help make it more effective.

Here are some general steps that can be taken:[10]

1. Define the Desired Culture: The first step in implementing a culture change is to define the desired culture. This requires a clear understanding of the organization's current culture and the changes that need to be made. The desired culture should be aligned with the organization's goals and values.

2. Communicate the Vision: Once the desired culture has been defined, it is important to communicate the vision to all stakeholders. This includes employees, customers, suppliers, and other partners. Communication should be clear, consistent, and ongoing.

3. Lead by Example: Changing organizational culture requires strong leadership. Leaders must set the tone by modeling the desired behaviors and values. This includes holding themselves accountable for upholding the new culture.

---

[10] See Hollister, R., Tecosky, K., Watkins, M., and Wolpert, C. (August 10, 2021).

4. Involve Employees: Culture change is more likely to be successful if employees participate in the process. This includes soliciting employee feedback, ideas, and input and providing opportunities for them to participate in the change process.

5. Provide Training and Support: Changing organizational culture often requires employees to learn new skills and behaviors. Training and support can help them adapt to the new culture more effectively.

6. Monitor Progress: Monitoring progress and adjusting the approach as needed are important. This includes tracking metrics related to the desired culture change and soliciting stakeholder feedback to ensure the transition has the desired impact.

Culture change is not a quick or easy process, but with a clear vision, strong leadership, and a commitment to involving employees and monitoring progress, it is possible to implement a successful culture change within an organization. The perspective presented in this book assumes that a mature data culture is an evolutionary journey that requires a culture change.

A Data Culture Perspective

Famed mathematician, philosopher, and inventor Charles Babbage once said, "Errors using inadequate data are much less than those using no data at all." This quote makes the case for a data culture that puts more data available at the point of decisions. The choice of organizational culture will directly impact the decision-making process. Having a data culture to support the organizational culture is vital.

What is a "data culture?" Data culture is the shared values, beliefs, attitudes, customs, and practices of an organization *in using data for decision-making* as they solve problems, reconcile dilemmas, and realize outcomes." With this foundation, organizations can understand their "data culture" and enhance the prospects of success with data projects such as master data management, metadata management, data governance, data warehousing/business intelligence, and analytics such as machine learning. Data culture encompasses the attitudes and beliefs about the role of data in the organization, the degree to which data is valued and trusted, and the extent to which data is used to drive business outcomes.

Data culture is characterized by several key elements, including a focus on data quality, a willingness to share data across the organization, a commitment to data privacy and security, a culture of continuous learning and improvement, and a recognition of the importance of data in achieving business objectives. It requires leadership support, employee engagement, and a strong data infrastructure to support data collection, analysis, and interpretation.

In today's rapidly evolving business landscape, organizations face the challenge of building a digital presence and leveraging data to drive growth, innovation, and operational excellence. Leaders strive for strategic advantage, informed decision-making, and a resilient organization. The key to achieving these goals requires cultivating a strong data culture—an organizational mindset and practices that prioritize the value and utilization of data.

Lasater, Albiladi, Davis, and Bengtson (2020) explored the use of data in Arkansas school districts in a multi-phased qualitative study. The researchers noted the differences between "data-driven decision-making" (DDDM) and "data-informed decision-making" (DIDM).[11] Data-informed decision-making includes "inquiry, analysis, and interpretation to inform decisions and actions" and not simply trigger decisions autonomously.[12] This differentiation allowed motive and purpose to be represented in their framework.

The researchers created a six-factor "data culture continuum." The continuum reflects the positive or negative data culture based on the six data factors. Each factor can be positive or negative, with some level of interdependency between them. The six data factors are as follows:

> Trust and Collaboration: The use of data for collaboration versus punitive measures. When trust is high, teachers do not mind sharing data among teachers and collaborating on opportunities for implementing better practices. When trust is low and data is used to embarrass or judge teachers, much less collaboration is likely.

---

[11] The process of interpreting data during decision-making. The interpretation of data can be based on many factors. Thus, the interpretation of the data may vary across constituencies.

[12] Lasater, Albiladi, Davis, and Bengtson (2020), page 535

Purpose of Data Use (Compliance Versus Improvement): Data is used for organizational learning versus accountability measurement. Teachers viewed a positive purpose of data use as improving student learning versus just measurement of outcomes to meet state, local, or national standards.

Leader Expectations and Teacher Agency: Leaders set the expectation that data was used, yielding teachers the latitude to make decisions freely with minimal oversight. Teachers viewed the data as a club when the expectation or systems established for data use were not clearly communicated.

Data Ownership: The teachers "owning" the data refers to the teachers' sense of responsibility for the results of the processes. This concept is heightened significantly by trust, collaboration, and shared ownership at the administration level. Dismissing the data because of correlations of other elements such as low income (free lunch) or parental difficulties is called "displacement" of data ownership.

Leadership Competency: Leadership (administrators, etc.) data competency, including demonstrating those competencies to the teachers by ensuring training of teachers, taking training themselves, and consistently using data in a competent and improvement-oriented manner.

Data As a Tool: Data used primarily for teacher evaluation with a slant toward improvement rather than just grading is positive.

In Florida, Brower et al. studied data culture from the perspective of distributing leadership and decision rights. The researchers surveyed over 600 Florida College System educators to determine the organization's data-sharing practices. The researchers identified a three-level continuum of data sharing: "need-to-know," democratic, and blended. "Need-to-know" culture was signified by low data sharing, lack of training, low collaboration, and reduced decision rights. In the democratic culture, leadership encourages sharing data, distributing leadership, collaborating, and training personnel to use data tools. In the blended environment, various practices were observed across the support, collaboration, access, and training vectors. These results demonstrate the need for understanding how leadership uses data to control decision-making and outcomes.

These two studies offer insight into the "competing values" in data culture. Using a similar model as Cameron and Quinn for organizational culture, two vectors were identified for describing data culture, which we also refer to as "competing values." Like organizational culture, these cultures are continuums and will display each vector to varying degrees. Likewise, data cultures may vary based on regional, departmental, or other criteria. As shown in Figure 1-2, a data culture crosses two continuum or competing values, which we refer to as "Usage" and "Flow."

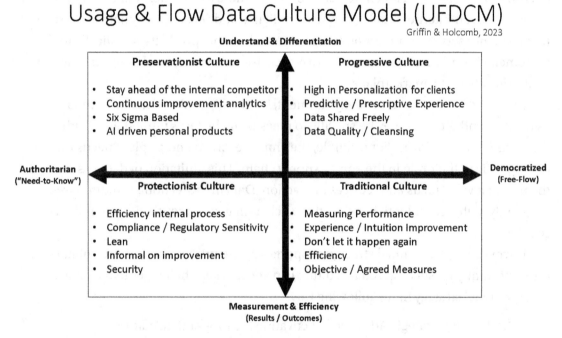

Figure 1-2. *The Usage and Flow Data Culture Model (UFDCM)*

The first is "the usage of data," such as measurement, diagnostics, and understanding. The other is the organization's enactment of the free "flow of data" from highly controlled to democratic. The result is a four-quadrant model representing different cultural types. The model is called the Usage and Flow Data Culture Model or UFDCM. Within data culture, two critical vectors shape an organization's approach to data: the Data Usage vector and the Data Flow vector. The Data Usage vector spans the purpose of Data Usage, from measurement/diagnostics to understanding/differentiation practices. In contrast, the Data Flow vector represents the continuum from authoritarian (need-to-know) to democratized access.

A strong data culture promotes a data-driven approach to decision-making, where data is used to inform and guide decision-making at all levels of the organization. It encourages a willingness to experiment, evaluate hypotheses, and learn from data, and it values the importance of collecting, analyzing, and interpreting data to gain insights and make informed decisions.

A Progressive data culture embraces a democratized approach to data, where accessibility and freedom of information are encouraged. This culture works well where decision rights can be easily shared and responsiveness to complex markets is necessary. For example, an Internet retailer may use predictive and prescriptive analytics to suggest that customer service representatives cross-sell additional products specifically to the customer. This information sharing between customer service, marketing, and sales is vital to loyalty in the marketplace.

In the case of a Protectionist environment, leadership makes data only available as needed, and the tools may be specific to measuring the tasks immediately within the focus of the employee. For example, the employee in an energy plant needs the data specific to their role in the energy supply chain. This limitation of data ensures the proper level of focus and reduces distraction. Data in Protectionist cultures are frequently highly confidential or sensitive, such as in government, utility, or military sectors.

Having an organizational strategy supported by the proper data culture enhances the opportunity for success. A data culture can address key challenges in realizing an organizational strategy in the following ways:

1. Driving Strategic Advantage: Activating the proper data culture empowers you to make strategic decisions based on reliable data insights, enabling you to identify market opportunities, anticipate customer needs, and stay ahead of the curve.

2. Enabling Informed Decision-Making: By embracing the proper level of democratized approach to data, you can foster a culture of data-driven decision-making, where employees have the necessary access and skills to utilize data effectively, leading to better-informed choices.

3. Fostering Innovation and Agility: Having the proper level of data flow to support the amount of experimentation, exploration, and data-driven problem-solving is vital.

4. Enhancing Organizational Resilience: Data governance and infrastructure are vital components of a data culture. By establishing robust data governance frameworks, you ensure data integrity, accessibility, and security. Additionally, investing in data infrastructure enables scalability and flexibility, providing a strong foundation for organizational resilience.

When an organization moves through the phases of the organizational life cycle, leaders must re-evaluate the data culture along with the organizational culture. With growth comes additional compliance, regulatory, and industry requirements, leading to a need to measure more precisely. Likewise, management must relinquish decision rights to more members of the organization. The data strategy must support this movement. At one point in the life cycle, the data culture may not work at other phases. Organizational and data culture are dynamic facilitators of organizational success.

Summary

This chapter examines the profound influence of organizational culture on the success or failure of strategic initiatives. Kicking off this examination is Peter Drucker's assertion that "culture eats strategy for breakfast." The chapter explores how an organization's underlying organizational and data culture shapes its ability to execute strategic plans effectively. The chapter continues with a definition of culture. In this context, culture refers to the shared values, beliefs, norms, and behaviors that define an organization. It encompasses employee attitudes, interdepartmental relationships, and the overall work environment. This definition is important as the exploration continues from organizational culture through to data culture.

The concept of a "data culture" is introduced as an extension of organizational culture. Data culture represents shared values, beliefs, attitudes, and practices related to using data for decision-making. It is vital for supporting data-driven initiatives and projects like master data management, data governance, and analytics. The chapter highlights key elements of a data culture, including data quality, privacy, continuous learning, and leadership support.

The Usage and Flow Data Culture Model (UFDCM) is introduced, representing the interplay between Data Usage and Data Flow. Four data culture types emerge: Progressive, Traditional, Protectionist, and Preservationist. Each type aligns with specific organizational needs, such as democratized data access for agility or controlled data access for security.

The chapter underscores the role of data culture in achieving strategic goals. Ultimately, the alignment of organizational culture and data culture is essential for realizing successful strategic outcomes throughout the organizational life cycle. In conclusion, the chapter emphasizes that just as culture shapes an organization's destiny, data culture plays a pivotal role in achieving strategic excellence. It highlights the dynamic relationship between organizational culture, data culture, and strategic success, calling for ongoing alignment and adaptation as an organization evolves.

CHAPTER 2

The Current State

Digital transformation spending is estimated to be $2.16 trillion in 2023, soaring to $3.4 trillion by 2026, according to Statista (2023). Organizations must understand who they are and what they want to be. English satirist Sir Terence Pratchett once wrote, "If you do not know where you come from, then you don't know where you are, and if you don't know where you are, then you don't know where you're going." Understanding how your organization reached its current state of organizational and data culture will give you an honest assessment of "where you are." Armed with understanding and a starting point, the migration from the current state to the desired state can be planned and executed. The future state organizational and data cultures must support each other to realize the organization's mission and satisfy environmental demands such as compliance, regulatory, or execution standards.

Assessing the current state of an organization's culture is a crucial step in building a mature data culture. By examining the organization through the lens of different cultural types and emphasizing the importance of data culture, we gain a comprehensive understanding of the current culture. This chapter explores how these cultural types shape our understanding of the current state and how they intersect with the components of a mature data culture.

To begin this chapter, we explore how to evaluate the current environment. First, we take an expanded view of the Usage and Flow Data Culture Model (UFDCM). In this view, the model explains the perspectives of each culture type with pros and cons. Finally, the expanded treatment will include some considerations when selecting a target culture type.

Next, we discuss several important components, such as organizational design and decision rights management, that can influence the flow of data. Leadership designs an organization to get the most work done with the least amount of friction. Most organizations employ multiple organizational designs to realize this goal. Furthermore, decision rights may be more restrictive in one area than in another based on the level of risk, compliance requirements, or tolerance for variance.

Finally, the components associated with a mature data culture are covered. A mature data culture in each quadrant of the UFDCM should implement these components. The differences vary based on the needs of the desired usage and flow of data. These components can lead to one or more programs that instill and stabilize the desired data culture.

Usage and Flow Data Culture Model (UFDCM)

As we evaluate the current state of an organization's culture, it is necessary to examine it through the lens of the different cultural types of Progressive, Traditional, Protectionist, and Preservationist, as described in the Usage and Flow Data Culture Model (UFDCM). These cultural types shape our understanding of an organization's current state. Each culture type possesses distinct values, beliefs, and approaches to change and development. However, it is also essential to emphasize the importance of building a data culture as we assess the current state. That is, how do these cultural types shape our understanding of the current state?

In the four quadrants of the UFDCM, the definition and focus for each illuminate their motivation for Data Usage (usage) and their characteristics for data sharing (flow). For example, a Progressive data culture will care about "understanding what is going on" and "the free flow of data throughout the organization." In contrast, the Traditional data culture will concern itself with "measurement and free flow." Both cultural types like the free flow of information, but their usage paradigms differ. In the Preservationist data culture type, an organization is focused on controlling the information, but the motivation is to know what they are doing to preserve or future-proof themselves. The data does not flow as freely inside or outside the organization, but they use advanced analytics to produce and use information to their benefit. A Preservationist data culture is great in a startup or anything high in intellectual property (IP) that needs to be protected but where data is being exploited to the extreme. A Protectionist data culture wants to control the information for measurement and only cares about repetition. A Protectionist culture fits the Six Sigma lean production space well. The data flows on a need-to-know basis when and where you need to know it.

With this overview, let's look at each of the four data cultures. These summaries discuss the level of Data Usage and Data Flow for each. Additionally, the pros and cons of the culture are elaborated. Each culture is a fit based on the need for innovation, repetition, security, and decision-making control.

1. Progressive Culture

    In a Progressive data culture, data freedom is high, allowing
    employees to access and share data freely within the organization.
    There is a democratized approach to data, where information is
    transparent, accessible, and readily available to all who need it.
    Data is seen as an asset for decision-making, and employees are
    encouraged to utilize data to gain insights and drive innovation.
    Data Usage in a Progressive culture emphasizes learning,
    experimentation, and data-driven decision-making. Innovation
    and continuous improvement are prioritized, and data literacy
    is nurtured to ensure employees can effectively use data in
    their roles.

    The pros of a Progressive data culture come from its free flow of
    data. When data flows freely, everyone can be informed about
    the organization broadly. This broadness allows the members to
    make more "data-driven" or better-informed decisions. When the
    organization's members have the right to act on the free-flowing
    data, the organization can become more adaptive to change.
    Furthermore, members of the organization may identify areas
    to help others throughout the ecosystem or to see where their
    function fits. This understanding of their place in producing goods
    and services can increase empathy for upstream and downstream
    and a desire to increase quality for those receiving outputs. When
    optimized, leadership is equipped with the information needed to
    perform strategic planning easily.

    Another pro of the Progressive culture is its use of data for
    understanding and differentiation. First, a Progressive culture
    readily adopts business intelligence and reporting tools. This
    adoption can include self-service capabilities to allow managers
    and potential front-line members to create data products such as
    reports or dashboards and unique analysis. In addition to these
    tools, predictive and prescriptive analytics, artificial intelligence,
    and other data exploitation techniques would be embraced. Bots
    to acquire new customers, assist current customers, or aid staff

in customer service environments are common tools. Finally, customer and employee personalization allows users to access data optimally for productivity.

With the free flow data and data tool proliferation, the amount of data can become overwhelming and distracting. First, leadership must be aware of "analysis paralysis" in decision-making. The abundance of information available can lead to overthinking or believing there is "one more fact" to consider that will perfect the decision. No scientific equation exists to aid leadership when "enough data" is "enough data," but being aware of the spiral of "one more datum point" can avoid paralysis in leadership decision-making.

Line workers and supervisors can become overwhelmed and distracted as well. Having many "slices" of data can confuse line workers. For example, an organization can have an "as-sold" view versus an "as-delivered" view versus a "revenue recognition" view. Each view can use similar labels, such as "revenue," yet have different results for a given time. This difference in results can lead to difficulty when line supervisors interact with other departments or units of the business.

For line workers, the proliferation of dashboards can be distracting as well. The balance of visualization to elementary data must be managed to ensure the line worker is not distracted. When locating the data rows to do their job, the line worker may need to traverse multiple visualizations, graphs of time series data, or drill levels. Although these representations can be helpful in planning, during operations, the line worker can become overwhelmed, distracted, and unproductive navigating them.

Leadership must consider the potential for internal misuse of data and tools in a Progressive culture. With the availability of both data and tools, the leadership must be aware that some users may attempt to create reports demonstrating a particular agenda. These "dueling numbers" can cost leadership time in debates and learning the rules. In a Progressive culture, the organization will constantly challenge the metrics it is using, and leadership must ensure this challenge is done in a productive and positive manner.

Beyond misuse, internal and external actors can inadvertently or intentionally exfiltrate data from the organization. The security surface is broad because data flow and tools are robust in a Progressive culture. Education and protective measures must be at the forefront of leadership priorities to ensure the optimization of the data asset and its security. The Progressive culture offers the highest levels of freedom and Data Usage for understanding for both the ethical and the nefarious actors.

2.   Traditional Culture

Like the Progressive data culture, the Traditional data culture values the free flow of data. Data is not restricted or tightly controlled in both cultures, and employees have access to relevant information. Data-driven decision-making, leveraging data insights to inform and support business decisions, is recognized as necessary for overall organizational performance by both as well. Both cultures acknowledge the significance of technology in managing and utilizing data effectively. Both invest in data infrastructure and analytics tools to enable efficient data management and analysis. As such, they strive for continuous improvement by using data insights to optimize processes, increase efficiency, and adapt to changing business needs and market conditions.

The primary difference between the Progressive data culture and the Traditional data culture lies in their level of Data Usage maturity along the "Data Usage" vector. The Progressive culture is positioned toward the "Understanding and Differentiation" end, emphasizing a more sophisticated and strategic approach to Data Usage. In contrast, the Traditional culture tends to be positioned closer to the "Measurement and Efficiency" end, with a focus on operational metrics and basic Data Usage. Of course, each organization may lie across the vector from (1) measurement and efficiency to (2) optimizing and improving to (3) strategic planning to (4) understanding and differentiation.

This difference can affect how the Progressive and Traditional data cultures execute strategic decision-making, innovation and differentiation, cross-functional interaction, and productivity. The Traditional data culture may use data to optimize existing processes and consider process management as strategic decision-making. Because efficiency and measurement are the focus, concepts of innovation and differentiation may be difficult in the Traditional data culture. Although data may be available, the focus on efficiency and measurement may cause members to become disinterested in data outside of their domain. This lack of interest can also lead to a lower level of cross-functional integration. Data will flow between organizations with limited depth of use because of a perceived lack of need. In a high-functioning Traditional culture, leadership takes steps to ensure a "systems" view so that interactions across the organization are understood and efficient. Even with these steps, the Traditional data culture will lag in proactivity compared to the Progressive data culture.

Traditional data culture has advantages over Progressive data culture as it seldom has difficulty with distraction because of the measurement and efficiency perspective. Although the data flows freely, the focus is more limited to measurement. In an organization where processes are stable and changing them can be costly, the Traditional data culture will be preferred. Likewise, when tolerance for variance is low, the data flow needs to be high, but measurement must be the focus. Using all the data to remove variance and increase efficiency to produce more successful outputs and fewer rejects is important.

The pros are similar for the Traditional data culture to the Progressive data culture based on the data flow vector. First, due to the high level of data availability, the organization will tend to make well-informed decisions. The decision in the Traditional data culture will be measurement and efficiency focused rather than understanding; however, both will be based on all the available data and considered well informed.

The cons based on the data flow vector are the same. Because data is available, security and misuse remain a concern for the Traditional data culture. Because the foci differ between the cultures, distraction should be lowered despite the data available. Likewise, because the data flows freely, the mission of efficiency and moving up the Data Usage vector should be easily facilitated with tools, processes, and leadership expansion of focus.

In the Data Usage vector, the pros differ. First, using data for measurement limits the amount of data necessary to operate. This reduction in the necessary data can translate into a focus on data quality and security on these data elements. With the focus on the reduction of variance in the processes, the data quality expectation will rise. Also, data security can focus on the most important data elements for advanced treatment. The measurement focus also reduces the likelihood that data will be possessed by individuals unless they are directly associated with the process. Having fewer users reduces the number of access paths for data, reducing the potential for incidental and intentional data exposures.

A Traditional data culture has several disadvantages to the Progressive data culture based on differences in the Data Usage vector. First, there is a lack of analytics being used in general. Although some analytics can be found in a Traditional data culture, they focus on measuring processes or outcomes. For example, artificial intelligence techniques can be employed for categorization to count outcomes. Traditional data culture includes descriptive analytics (what happened) and, based on where in the quadrant, can include diagnostic (why did it happen) and some predictive analytics (what is likely to happen). In comparison, Progressive data culture is characterized by always having descriptive, diagnostic, and predictive analytics with the addition of prescriptive (what should happen next) and cognitive analytics (what would a human think and do). Moving up the Data Usage vector is easiest when the data flows freely, as in the Traditional and Progressive data cultures.

3. Protectionist Culture

In a Protectionist data culture, data flow is low, and there is an emphasis on data security, data quality, and confidentiality. Data access is tightly controlled, and sharing may be limited to trusted partners or within strict boundaries. Data Usage in a Protectionist culture is often measurement oriented, focusing on using data to protect organizational interests, national security, and competitiveness. Decision-making prioritizes safeguarding against external risks and ensuring stability. Data privacy, data quality, and security are paramount concerns. Data is seen as a sensitive asset that requires strict protection.

The Protectionist data culture has the same profile as the Traditional data culture on the Data Usage vector. However, the Protectionist data culture explicitly controls the data in a more authoritarian manner on the data flow vector. This combination offers some unique pros and cons. Because the data flow is restricted through management decisions, the organization's exposure to external data loss is reduced. With less data and more targeted usage, the ability to secure the data, evaluate its quality, and regulate its uses is improved.

An obvious con in the Protectionist data culture is responsiveness to market dynamics. With limited data, the ability to perform advanced analytics may be limited. Likewise, limited data flow restricts the data for strategic planning, strategic decision-making, and cross-departmental interaction. Of course, if confidentiality and security are paramount to the organization's mission, then the Protectionist data culture is the perfect fit. For example, consider the use cases of governments and their associated contractors. The data flow must be limited to a "need-to-know" basis with high security and confidentiality.

4. Preservationist Culture

In a Preservationist data culture, data flow is low, focusing on internal preservation rather than external conservation. The organization emphasizes preserving its unique culture, values, and identity. Data is utilized with an understanding-oriented approach to gain insights into internal processes, performance, and improvement opportunities. The focus is on continuous improvement analytics and leveraging data to stay ahead of internal competitors. Preservationist organizations may adopt methodologies like Six Sigma to ensure efficiency and quality in their processes, and they may leverage advanced analytics, including AI, in manufacturing to optimize production and preserve their competitive edge. Data-driven decision-making is crucial to their preservation strategy, ensuring the organization remains adaptable and resilient in the face of internal challenges and market dynamics.

The Preservationist data culture offers the Data Usage of the Progressive data culture and the data flow of the Protectionist data culture. The data flows to resources based on the need to optimize the processes through analytical processing. When considering this type of culture, consider the needs of the medical or financial industry. The data in a medical center flow in a restrictive manner with high confidentiality but may go into complex models to determine a diagnosis or modify treatment. A similar situation exists for financial institutions as they process and evaluate transactions for regulatory reporting, money laundering, fraud, or other nefarious reasons.

In conclusion, the decisions to limit the flow and usage of data throughout the organization do not have simple rules. An organization may be in a highly regulated environment that demands it. Consideration of your industry or the industry your organization services may dominate this decision of cultural strategies. Rather, you may use different cultural paradigms based on organizational maturity (e.g., startup vs. Fortune 500), departments (e.g., marketing vs. operations), tasks (e.g., repetitive vs. custom), clients (e.g., government vs. public), regulatory requirements, or a combination of factors. The organization may leverage several micro-cultures within the organization for the flow of data and its usage based on need.

Organizational Structures

Organizational structures play an instrumental role in shaping the data culture within an organization. Building an organization to optimize decision-making and realize efficient production and the broadest opportunity for innovation comes with trade-offs. When assessing the alignment of organizational structures with the Usage and Flow Data Culture Model (UFDCM), we can examine how different structures influence data accessibility, collaboration, decision-making, and the overall data-driven mindset.

We explore several common horizontal and vertical organizational structures and how they affect data culture. Horizontal structures denote the segmentation of the organization at the highest levels. Vertical structures speak to the way an organization constructs levels in each segment. It should be noted that an organization will leverage multiple structures. An organization may be divided functionally (horizontal) and flat (vertical) because of its small size. Another organization may be divided divisionally (horizontal) and hierarchical (vertical). Larger organizations may use multiple horizontal and vertical strategies. For example, the organization may choose to use geography to segment horizontally with divisions (horizontal) and, within each division, have functional breakdowns (horizontal). Each functional area may be designed using a hierarchy (vertical) with teams of different configurations (vertical).

1. Functional Structure (Horizontal)

   The functional structure breaks up the organization based on the subject matter and expected outcomes. A functional structure may have organizations for sales, marketing, product development, operations, finance, etc. Most organizations have a functional aspect that requires purposeful deliberation when establishing data strategies and culture at the function and organizational levels.

   Each department in a functional structure has unique perspectives and frequently distinct personalities that can affect the cultural strategy. For example, the product development or research and development organization may seem glutinous data consumers but share little early in the product development or R&D process. They exhibit a Preservationist data culture because the amount of data they use is specific to the product but does not flow freely. Marketing may have a more democratized view of data to ensure sales, operations, and finance have marketing collateral, projections, promotional plans, etc. Legal and finance frequently limit data flow to maintain security and confidentiality. The data culture for such an organization can appear to be in all the quadrants.

What should be considered in the functional organization structure? First, ensure the data is flowing with common definitions and unique names. In a functional structure, marketing, finance, and sales can have different meanings for words like "customer" or "revenue." As previously mentioned, having different views such as the "as-sold revenue" view and "revenue recognition" views is legitimate; however, they can confuse employees when being compared. Depending on the organization, you may restrict the flow of some of these views into parts of the organization because you may deem those in that function do not have a "need-to-know." Conversely, you may decide to increase training to enlighten everyone and increase flow. Neither of these decisions is right or wrong; they are just decisions you should make to accommodate your organization.

Next, consider your outcomes. In the functional organizational structure, the outcomes are based on the function. Sales may be in farmer mode versus hunter mode in a mature marketplace. You have decided that Sales needs to attack new markets or add new products. These decisions on sales outcomes can change the dynamics of Data Usage and Data Flow. You may also employ advanced analytics when attacking new markets or considering new products. You would certainly train and educate sales reps on the data regarding the new products. This education could include different quantity concepts, new sales measurements, etc.

Finally, consider the current members of the organization. As mentioned, each function will have a unique perspective on the proper data culture. Some will be more conservative, advocating limited flow, and others more eager, advocating free flow. When you are making decisions to change the culture, consider these differences. Also, consider how long these individuals have been in the current culture. If you have a seasoned department, you may find cultural change more difficult than spinning up a new department. These members have changed over their tenure, and asking for an abrupt change can be difficult. This is especially true in a functional organization structure where the comfort with the culture has roots in the function's interests.

2.   Divisional Structure (Horizontal)

The divisional structure is characterized by breaking the organizations into "divisions" based on products, markets, geographies, or other alignment. For example, an organization may be broken into North American and European divisions. If the organization keeps all functions separated by division, then each division would be horizontally organized functionally. If the organization decided to break only the sales organization in the functional structure by geography, then only the sales division would have a divisional structure, and all others would operate as departments at the organizational level.

The divisional model has some important distinctions from the functional model. First, the division model allows organizations to account for national, industry, geographic, or other differences when crafting cultural strategies. Suppose the organization operates in an extremely sensitive industry such as national defense. In that case, the organization may create a "Government Division" to collaborate with those customers and their unique requirements, such as "FedRAMP" or security clearance. The "Consumer Division" for the same organization would work with nongovernment customers where security clearance is not needed. Obviously, Data Usage and Data Flow differ significantly between the Government Division and Consumer Division.

In addition to the differences within each division, Data Usage and Data Flow between divisions influence data culture strategy. In the previous example, the data flowing between the Government Division and Consumer Division will be limited. Other Data Usage and Data Flow considerations exist in other configurations, such as geographic divisions, for example, relative cost differences, currencies, national culture, and regulations, to name a few. Selecting the amount and types of data to be shared must consider these concepts. Likewise, some customer sets may have difficulty with analytics, such as neural networks, because they feel they are too opaque. Cultural alignment is vital to divisional structure success.

What should be considered in the divisional organization structure? Like the functional organization structure, common definitions and unique names are vital. These definitions are future complicated by the need to convert some metrics to account for different scales, for example, currencies. If sales revenue were not normalized to a common currency, a country like Mexico reporting in Pesos would appear to be a larger market than the United States. These conversions are vital when developing organization-level representations. Your cultural strategy may be to limit data crossing divisions to reduce the education and conversion burden.

The divisional structure adds new dimensions to the idea of the "customer." Customers may operate in multiple geographies or industries or be both governmental and consumer. In the divisional structure, the leadership must determine how to manage the data moving across the divisions to handle these customers properly. This alignment can be managed with either a data culture selection, limited or free data flow, or an organizational design of a team that oversees these "global" customers. In either case, the organization must determine how data will flow between the divisions and other parts of the organization.

Cultural strategy selection in the divisional organization structure allows for a more targeted application of Data Usage and Data Flow to match the division's characteristics. The opportunity to right-size data flow based on regulatory and compliance requirements, national cultural norms, or industry expectations offers a great opportunity for optimal Data Usage. Furthermore, the organization can easily isolate cultural nuances and reduce confusion and friction.

3. Matrix Structure (Vertical)

The matrix organizational structure emerges from the project needs of the organization. In this design, the employee reports to a functional manager and a project manager, creating a dual reporting relationship. The goal of the matrix structure is to leverage the expertise of functional departments while enabling flexibility and efficiency in managing projects. The functional manager, frequently referred to as the "resource" manager, cares for the subject matter expertise of the resources and their human resources administration, such as benefits, vacation planning, etc. The project manager manages the schedules and deliverables for the project team members. By incorporating cross-functional teams, this structure promotes collaboration and the sharing of data across different functional areas in the confines of the project.

A matrix structure brings together cross-functional teams for a short time to deliver a project or specific outcome. This structure promotes knowledge sharing to realize the project goals. For example, members from sales, marketing, IT, product development, and finance may work to launch a new product feature. The structure promotes the member to learn about the other functions' data. The matrix structure requires time for data literacy to ensure all team members are knowledgeable. Each new project team may consist of different members based on availability, so project and resource managers must factor in time for data knowledge acquisition.

The matrix structure requires monitoring to ensure a resource is not assigned to too many diverse projects. For this reason, resources are often assigned to the same domain and become domain subject matter experts. This specialization reduces the costs associated with training everyone on the data. By taking this approach, however, leadership reduces the benefits of data flow, which are possible in the matrix structure. The balance between specialization and cross-functional interactions for the entire resource pool will be based on the cultural strategy.

The matrix structure does not necessarily depend on the level of Data Usage as it would depend on the content of the project. The Data Usage vector may be measurement oriented, as in our example of the new product feature, unless the new product feature requires advanced analytics such as an anti-lock brake system, GPS, or autonomous driving automobile. In these latter cases, the use of data would be more extensive. Suppose the new product feature does not use data extensively. In that case, measurements such as trending and tracking the new feature's effects on sales, complaints, or profits may be negligible on the Data Usage vector.

4. Network Structure (Vertical)

The network and matrix structures are similar organizational designs that emphasize collaboration and interconnectedness across the organization. However, they have distinct differences in their approaches to authority, decision-making, and communication. Each difference has a distinct effect on how leadership manages Data Usage and Data Flow. Project complexity and specific expertise are key drivers in selecting a network structure. For example, an organization may leverage an outsourcing firm to cover specific expertise lacking in the organization.

As mentioned earlier, the matrix structure has a resource or functional manager and a project manager. In a network structure, authority and decision-making are decentralized.

The organization operates as a web of interconnected entities, often with no central authority. Each node in the network has a degree of autonomy and is responsible for its decisions. The entities collaborate based on shared objectives and trust rather than a hierarchical chain of command. This trust may be based on previous experience or contractual obligations.

Communication in a network structure is fluid and horizontal. Information flows freely among interconnected entities with less leadership intervention. There is no strict hierarchy dictating communication channels, fostering direct collaboration and knowledge sharing among all parties involved. This communication differs significantly from the matrix environment where the resource manager, project manager, and performer resources have hierarchical controls to ensure proper communication.

Based on the structure, data should flow freely within a network structure. Leadership must remain diligent so that data is shared at the correct level with the correct constituency organization. This diligence is especially needed with dealing with external entities. Data exfiltration and post-project data destruction are necessary to manage the data flow vector in a network structure.

Network structures are often adopted for complex projects or initiatives that require collaboration among diverse experts and entities. The flexible and collaborative nature of the network allows for the pooling of specialized skills and resources efficiently. Data Usage should naturally be higher as a project becomes more complex. The amount of data to monitor and analyze the interconnectedness of the project elements demands more usage techniques and combinations.

5. Team-Based Structure (Vertical)

The team-based structure, also called "self-managed teams," is like matrix structures. The self-managed team structure does not have a manager making daily decisions in the group. The team makes the decisions. Team-based structural design will have a person responsible for monitoring outcomes to ensure the team is producing the output. If output meets standards, the team has the autonomy to make decisions.

The matrix and team-based structures act similarly in project work. The project manager will be more active when a project structure is defined. In these cases, the two models work the same. For the remainder of this treatment, we will explore how data culture strategy will be associated with nonproject work in a team-based structure.

A team-based structure is defined by its association with a specific domain, such as a process or product. This approach limits the scope of the team and, thereby, the flow of the data. Because the team has specific outcomes, the data flows through the team to ensure those outcomes are met. Data outside the necessary data would distract from the production of those outcomes. The team may inquire regarding additional data flow to enhance outputs or improve processes. The team must seek this additional flow as management will only consider the performance based on its previously defined metrics.

The self-managed nature of the team-based structure requires the use of data for measurement. Depending on the expertise and tools available, teams may use advanced techniques to improve the process and increase outcomes. Team-based structures usually have metrics such as output per person or cycle. The team's focus and management's focus on measurement of the team based on outcomes limit the probability of advanced analytics being used in a team-based structure. Despite the low probability, teams augmenting their process with advanced analytics can realize gains and increase their value.

6. Flat Structure (Vertical)

    A flat organizational structure minimizes the number of hierarchical layers between employees and top-level management. Such a setup shortens the chain of command and decentralizes decision-making authority. The flat organizational structure empowers employees at various levels to make independent choices. A flat organizational structure promotes open communication, fosters a culture of collaboration, and encourages quick responses to dynamic market demands. Smaller organizations tend to be most suited for such a structure, especially in a startup environment.

A flat organizational structure allows more data to be shared across the organization. Because leadership and frontline employees have increased interactions, the amount of information flows easily. Because leadership works more directly with line workers, this design encourages data flow horizontally. The influence of frequent interaction with leadership will cause data to flow. Leadership must be aware that increased data flow can overwhelm and distract members. Getting the right level of data flow must be a focus of leadership. Increased data flow requires enhanced data literacy to optimize results and data security.

A flat organization's Data Usage will depend on the output of the organization rather than the structure. The flat organization promotes decentralized decision-making but does not make any demands on using data. As deeper insights are needed, the high frequency of interaction between leadership and line management should reduce barriers to implementing advanced Data Usage techniques.

7.   Hierarchical Structure (Vertical)

A hierarchical organizational structure is a traditional organizational design where authority, responsibility, and decision-making flow through a clear and structured chain of command. In this structure, employees are grouped into functional units, each with a designated leader responsible for managing their respective teams. Hierarchical organizational design is suitable when the organization seeks clear lines of authority, stability, and well-defined roles. It is commonly found in larger companies, government entities, and organizations with strict regulatory requirements.

Leaders select hierarchical designs throughout their organization to create explicit communication channels, areas of responsibility, and decision rights distribution. This explicit design dictates data flow along the specialization of the hierarchy. This specialization reduces Data Flow and can result in "data silos," making organizational analysis difficult. This structure frequently creates aggregations to enhance data flow through the levels. These aggregations must be managed well to ensure data flows accurately.

The hierarchical design promotes Data Usage for monitoring as leadership must hold middle management, line management, and line employees accountable. Unlike a flat organization design, top leadership in a hierarchical organization has less direct interaction with line workers and must rely on monitoring. The hierarchical design makes measurement throughout the hierarchy easy as the aggregation follows the chain of command and responsibility.

Organizations frequently create a functional design using hierarchical design within the functional area. Modern organizations are building groups to apply analytics to evaluate their operation and performance. These groups have expertise in tools and techniques to execute this analysis. The group collects data produced in the hierarchy, applies statistical analysis, and finds correlations and factors that can lead to improvements. This analysis requires a group outside the hierarchy due to the skills, tools, and objectivity needed for accurate analysis.

8. Virtual Work

Virtual work, also called "work from home," "telework," or "remote work," offers organizations access to expertise and bandwidth previously unavailable. An organization in West Virginia can find a database administrator in California rather than limiting the search locally. Virtual workers can exist in any of the organizational design structures. Virtual workers do not have the frequency of interaction with their peers, subordinates, and leadership.

Virtual work requires deliberate data flow. The lack of interactions with the virtual worker means data normally flowing though those interactions are not flowing. This data must be purposefully managed. Virtual workers report a sense of isolation, slower onboarding, and loss of nuances of change in processes. These losses result in restricted data flow. Leadership must explicitly put processes to ensure data flow to virtual workers to increase this data flow.

Virtual work allows further exploiting data for advanced analytics and enhanced usage. Data Usage for virtual workers performing operational work will be limited based on flow limitations. However, the opportunity exists to add skills in statistical modeling, analytics tools, or process improvement, allowing a team to advance Data Usage.

In summary, the structure will evolve much like everything else in the organization. In the parable of the three envelopes, a leader is given three envelopes by his retiring predecessor. The leader is instructed to open the envelopes one at a time when the going gets rocky. Following these instructions, the leader would open an envelope about every six to nine months. After opening and doing each recommendation, the environment would improve and then begin a negative slide. The first envelope had "blame the previous guy." The second envelope had "time to reorganize." The third envelope had "time to get three envelopes."

The second envelope had the only item of substance: reorganization. As an organization matures through the organizational life cycle or introduces new leadership, its structure will change. Implementing new strategies, attacking new markets, or replacing the retiring leader can be the catalyst for lasting change. The advancements in information and communications technology (ICT) create access to resource bandwidth and expertise previously unavailable. This access allows for new organizational structures. These new structures offer opportunities and challenges for establishing a healthy, mature, high-functioning data culture.

Components of a Mature Data Culture

While assessing the current state of an organization's data culture using the Usage and Flow Data Culture Model (UFDCM), it is important to consider how the model aligns with the components of a mature data culture. Here's an analysis of how the model can be applied to these components:

1. Data Strategy

   A data strategy aligns the "Data Usage" and "Data Flow" vectors. It establishes goals and objectives for managing and utilizing data effectively (Data Usage) while also addressing factors such as data quality, accessibility, and security (Data Flow). The data strategy accommodates the organization's strategic imperatives.

   A data strategy that emphasizes data democratization encourages a free flow of data across departments and teams. A data-driven organization with a democratized data flow empowers its workforce to make informed decisions at all levels. However, some environments require more data control. A data strategy that prioritizes control may restrict data flow to maintain data security, compliance, and cost control. As part of the strategy, leadership may decide to have more controlled data flow in parts of the organization than others. Leadership must consider the optimal flow to ensure decision-making is optimized.

   Data Usage in the data strategy includes plans for personalization, data analytics, data visualization, reporting, and data enrichment. Personalization can increase the value of each transaction by using data collected from customers. The spectrum of analytic (descriptive, diagnostic, predictive, prescriptive, and cognitive) and artificial intelligence offers insights to increase decision effectiveness. Visualization and reporting allow data to be presented, including trends, results, and other graphical representations. Data enrichment includes enhancing data through processing or combining it with other data like external data sources.

2. Data Governance Office

The data governance office (DGO) aligns most closely with the "Data Flow" vector of the UFDCM. A mature data culture requires a balance between accessibility and control over data. A robust DGO can foster a more controlled data flow by defining data access and usage policies. It ensures that data is managed efficiently and securely, mitigating potential risks associated with uncontrolled data democratization. With proper oversight, management can limit data access to specific teams or individuals, ensuring that only authorized personnel can access sensitive information while allowing flow. The DGO's responsibility for managing data governance ensures that data is accurate, complete, and secure, contributing to a higher level of Data Flow while maintaining necessary controls.

The DGO also plays an indirect effect on the Data Usage vector. Promoting best practices and ensuring compliance with data privacy regulations instill trust and confidence in Data Usage. A well-established DGO empowers employees with guidelines for proper Data Usage, safeguarding against misinterpretation and misuse of data. The DGO's effects on the data flow will ensure data is available for advanced Data Usage if so needed.

3. Data Governance Program

The establishment of a data governance program (DGP) is closely related to both the "Data Usage" and "Data Flow" vectors. The program defines and enforces policies and standards (Data Flow) while also focusing on effective and efficient Data Usage through data quality controls, classification, and retention policies (Data Usage). By establishing roles and responsibilities for data stewards, data custodians, and data users, the program can optimize the impact on Data Usage and Data Flow within the organization.

A well-executed DGP streamlines the flow of data by creating standardized data collection, storage, and distribution processes. Defining data ownership and stewardship ensures that data moves through the organization efficiently, facilitating access for those who require it while maintaining necessary controls.

The DGP enhances data literacy among employees by promoting a common understanding of data definitions and data quality reporting. The DGP also includes processes such as model validation to ensure data models are accurate and remain relevant. It ensures that data consumers have access to the right data at the right time, enabling better decision-making across the organization.

4. Clearly Documented Operational and Data Processes

Having clear and documented data processes supports the "Data Flow" vector by ensuring data accessibility and sharing within the organization. It also contributes to the "Data Usage" vector by promoting consistency and accuracy in data handling and utilization. The UFDCM recognizes that data is not a stand-alone entity but a product of an organization's processes.

By aligning data flow with the organization's key processes, management can ensure that data moves efficiently and meaningfully throughout the organization. Processes require data as inputs to produce their outcomes. Data is produced throughout the process steps. Once the outcome of the process is complete, data may flow out of the process to become an input to another process. Ensuring data accuracy and flow throughout the steps in the processes and to other processes is enhanced by having documented processes and the associated data flows. This alignment enhances data accuracy and relevance, promoting a controlled data flow supporting overall business objectives.

Process management helps establish a clear connection between Data Usage and organizational goals. Clearly documented processes with the associated data flows show the interconnectedness of processes and the process's data. By understanding how data relates to various processes, employees can better utilize data to optimize workflows and improve decision-making. Furthermore, when process changes are made, the cascading effects on data and potential additional usage opportunities can be easily identified.

5. Data Literacy

Data literacy in the context of the UFDCM refers to the education of employees on data meaning, handling processes, and tool usage. By promoting data literacy through training and resources, organizations empower employees to better understand and utilize data in their work. Data literacy supports a culture focused on understanding-oriented Data Usage.

From a data flow perspective, a workforce equipped with strong data literacy can contribute to a more democratized data flow. When employees understand the importance of data and its potential value, they are more likely to contribute to data-driven initiatives and share relevant insights. With data literacy, employees can identify opportunities for additional data flow that could lead to advanced Data Usage.

6. Data Analytics

Data analytics, including data visualization, data mining, and machine learning, falls within the realm of the "Data Usage" vector. A mature data culture leverages data analytics to gain insights and make better decisions, aligning with the understanding-oriented Data Usage emphasized in the model. Effective data analytics requires access to a diverse range of data sources. Therefore, data flow must strike a balance by providing access to relevant data while managing potential distractions that could arise from an overflow of information. Data governance measures can ensure that only the necessary data is made available to support analytical efforts.

7. Data Privacy and Security

Data privacy and security are crucial aspects of the "Data Flow" vector in the UFDCM. A mature data culture prioritizes protecting sensitive data through policies and procedures, ensuring that data is secure and accessible only to authorized individuals. Stringent data privacy and security measures may initially restrict data flow. However, when implemented effectively, these measures build trust among stakeholders, which can lead to more open data sharing and democratization in the long run. Clear protocols for data encryption, access controls, and user authentication can facilitate controlled data flow.

Data privacy and security directly impact Data Usage. With robust security measures in place, employees are more likely to feel confident in exploring data and conducting in-depth analyses. Additionally, adherence to data privacy regulations ensures that personalization efforts, when employed, respect individual data rights and preferences.

It's important to note that the UFDCM provides a framework for understanding the interplay between Data Usage and Data Flow within an organization. The components of a mature, high-functioning data culture can be aligned with this model to assess the organization's current state and identify areas for improvement.

Leadership must consider the cost and productivity effects of each initiative. First, cost consideration includes tools, processing, storage, security, and training based on the level of sophistication selected. For example, the cost of performing predictive analytics in real time may drive transaction costs too high, causing leadership to defer predictive analytics. Second, leadership must consider the productivity effects of each plan's implementation. For example, a line worker may become overwhelmed with visualizations designed by managers. This overwhelming feeling can lead to productivity loss and modifications to the visualization strategy.

In any assessment of your current environment, leaders will find various levels of adoption of each component. The adoption may be in some parts of the organization and not in others. The adoption may be at a lower level than desired. During the assessment, take honest stock of where these components are and the investments made to date. From that assessment, a strategy and comprehensive plan can be devised.

Charting Your Journey

When assessing the current state of an organization's culture and enacted programs and charting a path toward a mature, high-functioning data culture, the Usage and Flow Data Culture Model (UFDCM) provides a valuable perspective. The two vectors, Data Usage and Data Flow, allow leadership to focus on the needs of the organization and its functions. By examining the organization's phase in the organizational life cycle, the organizational structures with the associated decision rights, and the mission of the specific portion of the organization, a leader can determine the proper Data Usage and Data Flow levels. Implementing the optimal levels of usage and flow increases process efficacy, reduces distraction and frustration, and enhances decision-making throughout the organization. Components of a mature data culture, such as a well-defined data strategy and a data governance office, facilitate and sustain the desired culture. These components combine to regulate usage and flow based on the previously discussed needs of the organization and the desires of leadership. Here's an explanation of how these elements work together:

1. Assess the Organizational Life Cycle

    The organizational life cycle describes the phase of maturity of an organization. For our analysis, the organizational life cycle phases will include startup, growth, maturity, and decline/renewal. Leadership must evaluate the organizational structures, management team, organizational culture, decision rights, Data Usage, and data flow. Transitioning between phases can cause leadership to increase or decrease flow based on the organization's needs. In a startup, founders are the conduit for decision-making. As the organization scales, the founders can no longer be the only decision-makers. In the growth phase, data becomes vital to decision-making. The organization embraces a specialization in this phase, and the data flow reduces organization-wide. In the maturity phase, data exploitation is important to remain relevant in the market and defend market share. Without exploiting data for market relevance, the organization will enter the decline/renewal phase. In this phase, the organization can decide to rejuvenate its market presence, leading to advanced Data Usage. If the organization resigns to being end-of-life, Data Flow and Data Usage are restricted, and the business tries to get the most out of the assets and data.

Key questions to ask when evaluating an organization's life cycle phase:

- In what phase is your organization?

- Is the organization transitioning between phases?

- If in the decline/renewal phase, what is the vision (refresh or add new products, sell the organization, add new channels, ramp down, move to new markets, and others)?

- Has your organization had a major event (cyber-attack, regulatory/compliance change, funding infusion, new competitor, large contract, etc.) that may not cause a change phase but requires evaluation?

- From these answers, do you see additional needs for expanding Data Usage (analytics, reporting, visualization, personalization, artificial intelligence)?

2. Assess the Organizational Structures

As described earlier, organizational structures play a significant role in shaping the data culture within an organization. Consider both the horizontal and vertical structures of the organizational design. Increased levels in the organization's vertical structures restrict data flow. This restriction of flow must be monitored to ensure line workers have sufficient data to optimize their performance. The horizontal considerations differ based on the horizontal structure selected. In a functional structure, the interfunction data flow requires active training and data literacy. Data flow should be managed to encourage collaboration across different functions and moderate distraction. In a divisional structure, differences in process should be managed to ensure data is easily harmonized for leadership aggregations.

In addition to the structure, the decision rights must be assessed. For example, a hierarchy can share decision rights throughout the command chain. Senior leadership may retain decision rights at the highest levels or delegate to lower levels. Data must flow liberally in an environment where decision rights are delegated. Furthermore, Data Usage should be increased when decision rights are delegated to help less experienced employees make decisions. Leadership can equip personnel with tools such as inference engines, decision-tree expert systems, and artificial intelligence. These tools can also be used by middle and senior-level management to increase consistency and efficacy of decision-making.

Key questions to ask when evaluating organizational structure:

- How deep are the vertical levels in my current organizational design?

- Is the decision right delegated or centralized? Does data flow match the decision-right model?

- Are there groups within the organization where data must be restricted to

  - Remain compliant with regulations

  - Reduce distraction to maintain transaction throughput

  - Ensure confidentiality of intellectual property

- Are there areas of the organization where data flow needs to be increased to

  - Support advanced Data Usage such as customer personalization, analytics, new tools, or product development

  - Support current decision rights

  - Support processes and process-to-process interactions (reduce data re-entry)

- Are there areas or groups within the organization for increased Data Usage resulting in capabilities such as personalization (customer and employee), analytics, tool enhancement, product development, or others?

3. Assessing the Components of a Mature Data Culture

   Earlier in this chapter, seven components supporting the maturity of the selected data culture were presented. The current environment will have some or all these components established at some level. Interestingly, each component must be evaluated in the current environment. Some questions for each component are as follows:

   - Data Strategy

     - Does your organization have a comprehensive data strategy?

     - Does the data strategy have principles well understood by employees?

     - Do employees know where to go if they have questions about the data strategy?

   - Data Governance Office

     - Do you have a data governance office?

     - Do the employees of the organization know the data governance office exists and its mission?

     - What level of autonomy does the data governance office have?

   - Data Governance Program

     - Does your data governance program have elements based on Data Usage and Data Flow?

     - Is your data governance program well promoted?

   - Clearly Documented Organizational and Data Processes

     - Do you have organizational processes documented? Are there processes to maintain the process maps?

- Do you have the data flows associated with organizational processes documented?

- Do you have the data flows associated with the analytical and reporting processes documented? Are there processes to maintain the process maps?

- Does the organization have a data dictionary for all data elements and processes? Are there processes to maintain the data descriptions?

- Data Literacy

  - Does your organization have formal time allotted for learning data access and data meanings during

    - New employee onboarding

    - Employee transfers, promotions, and job changes

    - Application or tool introductions and upgrades

    - Process changes

    - Adding new providers or partners

    - New products or services

  - Does your organization have formalized training for Data Usage rules, including those for compliance and regulatory requirements?

  - Does an employee know who to contact with questions regarding data literacy needs?

- Data Analytics

  - What descriptive analytics (including reporting, dashboards, etc.) does your organization perform?

    - What tools are used?

    - Is the process to initiate the creation of the descriptive analytics products (like reports and dashboards) well documented and understood?

    - Are the usage, efficacy, and benefits of these products measured?

- What diagnostic analytics and analysis does your organization perform?

  - What tools are used?

  - Is the process to initiate the creation of the diagnostic analytics products well documented and understood?

  - Are the usage, efficacy, and benefits of these products measured?

- What predictive analytics and analysis does your organization perform?

  - What tools are used?

  - Is the process to initiate the creation of the predictive analytics products well documented and understood?

  - Are the usage, efficacy, and benefits of these products measured?

- What prescriptive analytics and analysis does your organization perform?

  - What tools are used?

  - Is the process to initiate the creation of the prescriptive analytics products well documented and understood?

  - Are the usage, efficacy, and benefits of these products measured?

- What cognitive analytics, machine learning, and other artificial intelligence and analysis does your organization perform?

  - What tools are used?

  - Is the process to initiate the creation of these advanced analytics products well documented and understood?

  - Are the usage, efficacy, and benefits of these products measured?

- Data Privacy and Security

  - Does your organization have a Chief Information Security Officer?

- Does your organization have documented processes for handling data to ensure confidentiality, accessibility, and integrity?

- Does your organization actively monitor data for secure usage and flow?

- Does your organization have data privacy and security requirements and sign-off for all projects?

By combining the assessment of the components of a mature data culture with the analysis of organizational structures and organizational life cycle, organizations can identify gaps and areas for improvement in their data culture. This analysis helps organizations understand where they currently stand and what needs to be addressed to build a mature data culture. For instance, if the assessment reveals a lack of a dedicated data governance office or clear data processes, the organization can prioritize establishing these structures to ensure proper data management and governance. Suppose the analysis shows that the current organizational structure hinders data accessibility and collaboration. In that case, organizations can explore options for adopting alternative structures, such as a matrix structure or a team-based structure.

4. Developing a Strategic Road Map

Based on the assessment and analysis, organizations can develop a strategic road map for building a mature data culture. This road map outlines the specific actions, initiatives, and milestones required to bridge the gaps identified and align the organization with the desired data culture. The road map may include initiatives such as developing a comprehensive data strategy, establishing a data governance office, implementing a data governance program, documenting data processes, developing a comprehensive data strategy, investing in data literacy programs, enhancing data analytics capabilities, and strengthening data privacy and security measures. The road map should also consider any necessary adjustments to the organizational structure to support the desired data culture.

By following the strategic road map, organizations can systematically address the identified gaps. Each step will ensure the usage and flow are optimized to realize the data culture that matches the organizational decision-making culture, structure, and phase in the organizational life cycle. The assessment process will identify positives you can build upon and areas needing work. Implementing each step thoughtfully will gradually transform the culture into a mature data culture that values data accessibility, collaboration, data-driven decision-making, and effective data governance.

Summary

This chapter begins with an expanded view of the Usage and Flow Data Culture Model (UFDCM) and its four resulting culture types: Progressive, Traditional, Protectionist, and Preservationist. These cultural types influence an organization's current state and shape its values, beliefs, and Data Usage and Data Flow approach. Each type's motivation for Data Usage and characteristics of data sharing are illuminated, showcasing their impact on decision-making, innovation, security, and efficiency. The pros and cons of each culture type consider factors such as data availability, security, distraction, and alignment with organizational goals. It underscores the need to align data culture with organizational needs, industry, regulatory requirements, and other factors, suggesting that a combination of cultural paradigms can be adopted based on specific contexts within the organization.

After a deeper dive into the UFDCM, the impact of an organization's design and structure on developing a data culture is explored. These structures influence data accessibility, collaboration, decision-making, and a data-driven mindset. Different structures lead to varied Data Flow and Data Usage patterns, affecting departments' perspectives and cultural strategies. Considerations for data alignment, outcomes, member perspectives, and complexity are discussed within each structure. The organizational structure evolves and changes over time, influenced by factors such as leadership changes, market strategies, and technological advancements, which all play a role in shaping a successful data culture.

Next, the components of a mature data culture are the starting point for building a data culture within an organization. The UFDCM leverages these components to establish and sustain the desired data culture. The components include data strategy, data governance office (DGO), data governance program (DGP), clearly documented operational and data processes, data literacy, data analytics, and data privacy and security. Each component contributes to Data Flow and Data Usage in unique ways and must be balanced for success.

Finally, leaders must understand the current culture and progress on the components of a mature data culture. Questions for the leader to evaluate their current culture and implementation of the components of a mature data culture are offered. The questions offer a guide to understanding the starting place for the organization. With this knowledge, initiatives can be designed to move toward the desired data culture.

CHAPTER 3

Organizational Vision and Data Strategy

The Art of War author Sun Tzu said, "Strategy without tactics is the slowest route to victory. Tactics without strategy is the noise before defeat." This adage applies equally to business. Leadership must craft strong strategies with solid tactics to realize victory. In this chapter, we begin the discussion of creating the data strategy with an eye toward tactics. We emphasize that recognizing organizational culture is the starting point for creating a data strategy. The exact data strategy needed for a particular organization can be crafted by examining the current data culture through the Usage and Flow Data Culture Model (UFDCM). Like all key components of a mature data culture, data strategy is not a one-size-fits-all. Identifying the appropriate quadrant in the UFDCM for an organization sets the tone and direction of the data strategy.

Data strategy is a plan that outlines how an organization will collect, manage, store, share, and use data to achieve its business goals. It is a comprehensive approach to managing data as a strategic asset that can provide insights and competitive advantages. A good data strategy is key to achieving a successful data transformation. Without it, it's extremely difficult, if not entirely impossible, to know exactly where you're at on your transformation road map.

A well-designed data strategy can help organizations to improve their operations, enhance customer experiences, and drive innovation and growth. A data strategy typically includes the following components:

1. Business Objectives: A data strategy should align with the organization's overall business objectives and support its mission, vision, and values.

2. Data Governance: A data strategy should establish policies and procedures for data quality, security, privacy, compliance, and accessibility.

51

G. W. Griffin and D. Holcomb, *Building a Data Culture*, https://doi.org/10.1007/978-1-4842-9966-1_3

3.   Data Architecture: A data strategy should define the types of data that the organization will collect, the systems and technologies that will be used to manage and store the data, and the processes for data integration and analysis.

4.   Data Management: A data strategy should outline how the organization will manage data throughout its life cycle, including data acquisition, cleansing, storage, retrieval, and disposal.

5.   Data Analytics: A data strategy should define the methods and tools that will be used to extract insights from data and inform decision-making.

6.   Data Culture: A data strategy should foster a culture of data-driven decision-making, where employees are trained and empowered to use data effectively in their work.

This chapter examines data strategy in each of the quadrants of the Usage and Flow Data Culture Model (UFDCM). A discussion of data strategy is presented in the context of organizational strategy. The data strategy must support the organizational strategy. This support may include transformation and change that meet resistance and as such must be addressed. Techniques for addressing this cultural resistance are presented. Establishing and sustaining a data culture will likely require new skills and attitudes, so a section is offered on talent development. Finally, a look at nuances and differences of data strategy in each quadrant is presented.

Why Do We Need a Data Strategy?

A data strategy plays a crucial role in the modern, data-rich business environment. It serves as a comprehensive road map that outlines how an organization will effectively collect, manage, store, share, and utilize data to achieve its business objectives. The need for a data strategy has become even more apparent due to the increasing volume, complexity, and variety of data available to organizations and the rapid advancements in data technologies.

According to Edwards (2023), seven major trends are reshaping organizations' current data strategies, including the adoption of real-time data and complexity management, facilitating in-house data access, encouraging external data sharing, embracing data fabric and data mesh technologies, prioritizing data observability, treating data as a product, and forming cross-functional data product teams. These trends reflect the growing importance of data-driven decision-making and the constant need for organizations to evolve their data strategies to keep up with technological advancements and business demands.

The importance of having a well-defined data strategy is discussed in the article titled "Data Strategy Definition: 7 Key Elements of Data Strategy" (MasterClass, 2022). It emphasizes that a data strategy serves as a road map to align data management with business goals, and it involves critical elements such as understanding organizational strategy, optimizing data access, deciding data architecture, integrating data, implementing data management processes, identifying data sources, and transforming data into actionable insights. A well-executed data strategy leads to increased productivity, improved business reputation, enhanced data quality, better regulatory compliance, and informed decision-making.

The data strategy must fit the organizational strategy. The article "Data Strategy to Manage Data Successfully" (CDQ, n.d.) emphasizes that data strategy should align with the organizational strategy to optimize data utilization and overcome challenges such as data silos, data defects, and legal and regulatory requirements. Without this alignment, the data strategy will conflict with the goals of the organizational strategy, resulting in frustration, missed opportunity, and failure of both.

A Chief Data Officer (CDO) can increase the data strategy's success. The article "How to develop an enterprise data strategy" (IBM Cognitive, n.d.) discusses the role of a CDO in developing and executing an enterprise data strategy. The authors emphasize the importance of data governance, consolidation, and training of data specialists. The CDO should have both the business and technical acumen to ensure the fit of the data strategy to the organization. The CDO can right-size the data strategy with an eye on the transformative potential of cognitive computing and suggest catalyst projects to accelerate the adoption of cognitive technologies.

The data strategy must consider freedom of data flow throughout the organization. Tekiner and Bak (2023) stress the importance of balancing data defense and data offense to derive value from data. Organizations possessing high-valued assets or operating in highly regulated industries tend more toward data defense and restricted flow. Organizations in more customer-facing industries, such as Internet retail, tend toward data offense. A well-balanced data strategy can lead to significant cost savings, better decision-making, and improved competitiveness.

The data strategy must be turned into tactics, projects, and initiatives to be realized. Bernard Marr's book on data strategy highlights the reasons behind data strategy failures. In Chapter 11, Marr (2017) emphasizes the importance of effective communication, buy-in from all levels of the organization, and a collaborative relationship between data functions and business leaders for successful strategy execution. Additionally, the data strategy implementation requires proper program and project management, clear objectives, and well-defined metrics. Managing data projects requires a balance of process, business, and technical acumen. Selecting these skills in project managers and having the proper leadership of a qualified Chief Data Officer can enhance success.

Beyond the data strategy's content, the data strategy acts as a communication vehicle for the organization's current and desired data state. The data strategy communicates the organization's intentions and plans to exploit its data in operational, tactical, strategic, and aspirational ways. Leaders must support the initiatives, be personally engaged, and offer positive reinforcement. The communication of intent coupled with engaged leaders and the elements of business objectives alignment, robust data governance, efficient data architecture, effective data management, data analytics capabilities, and a data-driven culture equip leadership, especially the CDO, to realize the optimal data culture (Edwards, 2023; MasterClass, 2022; Marr, 2017; CDQ, n.d.; IBM Cognitive, n.d.; Tekiner & Bak, March 23, 2023).

Data Strategy and Organizational Strategy

A data strategy must align with the organizational strategy. The two are interconnected aspects that should be aligned to achieve a company's goals and objectives. In today's digital world, data has become the lifeblood of businesses. When decision-makers possess data about internal strengths and weaknesses and external opportunities and threats as well as environmental elements such as the political, economic, social, technological, ecological (environmental), and legal (regulatory) dynamics, decisions become more adaptive. These data-driven decisions optimize people, processes, and technology. Without alignment, error rates in processes, frustration of employees, and underutilization of technology can result.

First, aligning data strategy with organizational strategy helps companies identify the data that is important to the business. Evaluating each objective and goal of the company with the data needs will ensure proper flow. The two strategies converge through the processes supporting each objective and goal. Organizational strategy defines the objectives and goals of a company, and data strategy outlines how data can support those objectives. As processes are identified from the organizational strategy, the well-crafted data strategy will naturally accommodate the data needs. This convergence allows leadership to ensure the proper data flows to the decision-makers at the time of the decision.

Second, aligning data strategy with organizational strategy helps maximize the data's value. Without a clear understanding of how data can support business goals, companies may collect and analyze data that is not relevant or useful. This wastes people and technology resources and can lead to wrong decisions and bad processes. By aligning data strategy with organizational strategy, companies can ensure that they collect and analyze the right data, which helps make informed decisions. This, in turn, maximizes the value of data and helps companies gain a competitive advantage.

Third, aligning data strategy with organizational strategy helps create a data-driven culture. When data strategy is aligned with organizational strategy, it becomes an integral part of the company's decision-making process. During organizational strategy development, leadership will determine who has the power to make decisions. Empowering people to make decisions is not enough; they must be equipped with the tools, specifically the data, to be successful. The data strategy will create these tools for decision-makers. This helps in creating a data-driven culture, where decisions are made based on data and not just intuition. This culture encourages employees to use data in their decision-making processes, which can lead to better outcomes.

In addition to tactical decision-making improvements, aligning data strategy with organizational strategy helps identify opportunities for innovation. In the organizational strategy, the organization will plan the development of processes and products. The data strategy will include structures such as data stores, analytics tools, and data analysts. These resources can be used to uncover opportunities for market, product, process, or resource innovation. For example, by analyzing customer data, companies can identify new products or services that customers may want. This can lead to new revenue streams and growth opportunities for the company.

Data strategy and organizational strategy are interdependent. Aligning data strategy with organizational strategy helps companies identify the right data, maximize the value of data, create a data-driven culture, and identify opportunities for innovation. Therefore, it is essential for companies to ensure that their data strategy is aligned with their organizational strategy to achieve their goals and objectives.

Data Strategy and Data Culture Interaction

The relationship between data strategy and data culture can significantly impact an organization's success in leveraging data effectively. A well-crafted data strategy must consider the prevailing data culture within the organization, and in turn, a data culture influences and shapes a data strategy that aligns with organizational strategy objectives. Culture is the starting point for understanding and crafting a successful data strategy. The Usage and Flow Data Culture Model identifies four distinct data culture types: Progressive, Traditional, Preservationist, and Protectionist. Each data culture type has its unique characteristics and attitudes toward Data Usage and Data Flow.

For instance, in a Traditional data culture, decision-making is often based on historical data and intuition, with little emphasis on data-driven insights beyond diagnostics or transactional throughput improvements. In contrast, a Progressive data culture embraces data analytics to understand and differentiate through analytics, personalization, etc. Each culture has its pros and cons, as mentioned earlier. Traditional cultures can miss opportunities that advanced analytics could illuminate. A Progressive culture can overwhelm areas of the organization or become over-reliant on data and miss opportunities for advancement. Understanding these data culture types allows organizations to tailor their data strategies to address specific challenges and opportunities within their unique organization.

The data strategy must promote cultural change and alignment to ensure successful adoption and implementation. For instance, in a Protectionist data culture, where data is closely guarded and access is restricted, the data strategy should focus on building trust and transparency around Data Usage to encourage broader data access and sharing. Organizations that correctly align their data strategy with their data culture drive positive outcomes. The article "Data Strategy Definition: 7 Key Elements of Data Strategy" emphasizes the importance of a data strategy in supporting business goals and outlines key elements such as data access, data architecture, and data analytics. It highlights that when organizations successfully implement data strategies that align with their specific data culture, they are much more successful. By recognizing the interplay between data strategy and data culture, organizations can foster a data-driven culture that empowers employees to make data-informed decisions. This highlights the significance of engaging key personnel in data strategy development to create data advocates within the organization. These advocates play a crucial role in promoting data-driven decision-making and building a data culture from the top leadership to every layer of the organization.

The relationship between data strategy and data culture is symbiotic. A well-designed data strategy aligns with the prevailing data culture and, at the same time, sets the stage to continue to make the shift required to the desired data culture. By viewing data strategy through the UFDCM lens, organizations can learn and adapt their data strategies with a well-defined target culture. This strategy and planning will drive positive cultural change, leading to better data utilization, improved decision-making, and competitive advantages.

Overcoming Challenges in Data Culture Transformation

In the data strategy, leadership will determine if the current data culture will be the future data culture. This target data culture may restrict data flow to centralize decision rights or expand data flow to support delegated decision rights. Data flow could be increased to allow better process management or customer self-service. Leadership may also decide to invest in prescriptive analytics to make processes better or cognitive analytics to do customer simulations. These changes in the usage and flow vectors will change the data culture. This section explores the challenges that organizations may encounter when transitioning from one data culture type to another and offers strategies for overcoming resistance to change. By addressing common barriers and providing practical advice, organizations can successfully foster a data-driven mindset across different levels of the organization.

One of the primary challenges in data culture transformation is cultural resistance. Cultures resist changing because it requires thinking and acting differently. Before the culture was Preservationist, it moved from being one of the other three. In Traditional or Protectionist data culture, employees may be accustomed to decision-making based on limited data. Frequently, these cultures have centralized decision rights. Convincing employees to embrace increased Data Usage capabilities such as advanced analytics and to trust the resulting insights can be challenging. Furthermore, leadership must embrace the change and give up decision-making based only on their experience and intuition.

Similarly, in organizations with a Protectionist and Preservationist data culture, where data flow has been closely guarded, breaking down these barriers and promoting a collaborative data-sharing environment may meet with reluctance and resistance.

In these cultures, data literacy levels may be lower, leading to a lack of understanding of the breadth of data that could be available and its potential and limitations. Addressing this challenge requires investing in data literacy training programs to empower employees to interpret and utilize data effectively.

Securing leadership buy-in and support is crucial for data culture transformation. Leaders must champion the shift toward data-driven decision-making and function as role models by incorporating data insights into their own decision-making processes. As previously mentioned, leaders must be ready to relinquish previously held decision rights or actively take decision rights back to realize the target culture. In each case, leaders must be actively involved and positively moving toward the target data-driven decision-making culture.

To overcome resistance, organizations must communicate the benefits of a data-driven culture to employees at all levels. Demonstrating how data-driven decision-making can lead to improved outcomes, enhanced efficiency, and better performance can motivate employees to embrace the change. Investing in data literacy training programs is essential for improving employees' understanding and confidence in using data. This training can range from basic data literacy courses to advanced data analytics and visualization workshops.

Data governance also plays an important role in data strategy execution, regardless of the data culture type. Chapter 5 provides a much more extensive explanation of data governance using the UFDCM.[1] By establishing a robust data governance framework and policies that align with the characteristics of each data culture, organizations can ensure data quality, compliance, security, and privacy while maximizing data utilization. A well-executed data governance strategy paves the way for successful data-driven decision-making, regardless of the organization's data culture type.

Technology adoption is another key in supporting data strategy execution across different data culture types. Selecting and implementing appropriate technologies that align with the organization's data culture enhances data-driven decision-making, fosters innovation, and maximizes the value of data assets. By understanding the technology

---

[1] This section is not meant to fully examine data governance. Chapter 5 provides readers with a much more extensive explanation of data governance using the UFDCM.

needs and priorities of their data culture, organizations can build a technology stack that complements their data strategy and empowers employees to leverage data effectively. When moving from one data culture type to another, leaders must be especially watchful for technology adoption issues.

Overcoming challenges in data culture transformation requires a multifaceted approach that includes leadership buy-in, effective communication, right-sized data literacy training, and optimized data sharing. By incorporating real-world examples and case studies, organizations can learn from successful transformational efforts and adopt strategies that align with their current and desired data culture types. A successful data culture transformation empowers employees to use data effectively, leading to improved decision-making, increased efficiency, and a competitive advantage in the data-driven business landscape.

Data Strategy and Talent Development

The meme saying "Teamwork makes the dream work" is true. When an organization invests in its people, change comes much more easily. When it comes to data culture, the data strategy needs to account for the talent to advance to the desired data culture. This section highlights the importance of talent development, training, and upskilling programs to ensure that employees are equipped to leverage data effectively. In some cases, talent augmentation will be necessary as well. By fostering a culture of continuous learning and data literacy, organizations can empower employees in all data culture types to make data-driven decisions and contribute to the overall success of the data strategy.

A well-designed data strategy relies on employees' ability to effectively collect, analyze, and interpret data. Investing in talent development ensures that employees have the necessary skills to leverage data to its fullest potential. Data-driven organizations encourage employees to explore and analyze data creatively. Talent development programs can foster innovation and allow employees to discover insights that drive business growth and innovation.

As technology evolves, employees need to stay updated with the latest data tools and analytics platforms. Talent development ensures that employees are well equipped to adapt to technological advancements and use them to their advantage. Implementing data literacy training programs is essential for all data culture types. These programs should cover basic data concepts, data visualization, data analysis techniques, and data ethics, enabling employees to comprehend and communicate data effectively.

Upskilling employees in advanced data analytics, machine learning, and artificial intelligence can lead to a more data-savvy workforce. Organizations can offer workshops, certifications, and mentorship programs to foster employees' proficiency in these areas. Promoting knowledge sharing among employees encourages a collaborative learning environment. Creating forums, internal communities, or data-focused events allows employees to exchange best practices and share data insights in any data culture.

A data strategy can only be successful if employees have the necessary talent and skills to leverage data effectively. Talent development programs play a crucial role in fostering a data-driven culture within an organization. By investing in data literacy training, upskilling initiatives, and knowledge-sharing platforms, organizations can empower employees to make data-driven decisions regardless of the data culture type. A well-equipped and data-savvy workforce is the foundation of a successful data strategy, leading to better decision-making, improved operational efficiency, and enhanced innovation.

A Survey of Data Culture Within Each Quadrant

The development of a data strategy must consider the current data culture and the target data culture. The data strategy will outline the use of the various components to move the culture to the desired state. This maturation will shift an organization's placement on the two vectors of Data Usage and Data Flow. The Data Usage vector can shift through the continuum of measurement, diagnostics, understanding, and differentiation. The Data Usage vector is an "additive" continuum, meaning as an organization adds Data Usage levels, the other capabilities are included. For example, a data culture on the differentiation end of the continuum also employs measurement, diagnostics, and understanding. The data flow vector can shift across the continuum of limited data flow to a democratized or free data flow.

This section examines what a data strategy may look like for each data culture type. Within an organization, characteristics of all data culture types may exist based on a specific department's function. Selecting the right type of data strategy to manage these dichotomies can be challenging. The UFDCM is designed to guide the practitioner in managing the complexities of the entire organization's data culture needs around data strategy.

Data Strategy in a Preservationist Culture

In a Preservationist culture, where data flow is controlled and Data Usage includes measurement, diagnostics, understanding, and differentiation, organizations place significant emphasis on data preservation and security. While access to data is restricted, the organization recognizes the strategic value of utilizing data for decision-making. This cultural perspective can be seen in the medical industry, where data is highly controlled and state-of-the-art models leverage that data to diagnose and treat patients. In this section, we explore what data strategy looks like within the context of a Preservationist culture using the Usage and Flow Data Culture Model (UFDCM).

1. Data Preservation and Security

    Data strategy in a Preservationist culture revolves around preserving the integrity and security of data assets. The organization implements stringent data protection measures, including robust encryption, backup and recovery systems, and access controls. It prioritizes the establishment of strong data governance frameworks and compliance with relevant data protection regulations.

2. Controlled Data Utilization

    While data accessibility is limited, data strategy in a Preservationist culture focuses on controlled data flow. The organization establishes clear guidelines and protocols for authorized individuals or teams to access and utilize data. It ensures that only those with the necessary permissions and expertise can leverage data for decision-making and strategic initiatives.

3. Secure Data Sharing

    In a Preservationist culture, data strategy recognizes the need for secure data sharing. The organization establishes mechanisms and platforms for securely sharing data with external partners or stakeholders when necessary. This may involve implementing secure data exchange protocols, data anonymization techniques, and contractual agreements to safeguard data confidentiality.

4. Preservation of Data Privacy

    Preservationist data strategy places a strong emphasis on data
    privacy. To protect sensitive information, the organization
    implements rigorous privacy measures, such as data masking, de-
    identification, and data access audits. It ensures compliance with
    privacy regulations and cultivates a culture of privacy awareness
    among employees.

5. Data-Driven Decision-Making Within Boundaries

    Data strategy in a Preservationist culture enables data-driven
    decision-making within the boundaries of data security and
    access restrictions. The organization invests in building internal
    data analytics capabilities, ensuring that authorized teams have
    the necessary skills and tools to extract insights from available
    data sources. This allows for evidence-based decision-making
    while adhering to the Preservationist principles.

Data strategy must strike a delicate balance between data utilization and security in
a Preservationist culture. By prioritizing data preservation and security, implementing
controlled data utilization practices, enabling secure data sharing, preserving data
privacy, and fostering data-driven decision-making within established boundaries,
organizations can leverage their data assets while maintaining a cautious approach to
data accessibility. A well-executed data strategy in a Preservationist culture enables
organizations to protect valuable data while still deriving insights for informed
decision-making.

Data Strategy in a Protectionist Culture

In a Protectionist culture, where data flow is controlled and Data Usage is limited to
measurement and diagnostics, organizations tend to restrict access to data and tightly
control information flow. However, a well-designed data strategy can help optimize
data-driven decisions while addressing data protection and governance concerns. In
this section, we explore what data strategy looks like within the context of a Protectionist
culture using the UFDCM.

1. Establishing Data Governance

   Data strategy in a Protectionist culture starts with a strong
   emphasis on data governance. The organization recognizes the
   importance of protecting sensitive data assets and establishes
   robust policies and procedures to ensure data security, privacy,
   and compliance. This includes defining access controls, data
   classification, and data retention policies and establishing data
   stewardship roles. The data governance in a Protectionist and
   Preservationist culture will be very similar in content; however,
   the Protectionist culture is generally seen in the most sensitive
   environment such as governments.

2. Controlled Data Accessibility

   In a Protectionist culture, data strategy focuses on controlled data
   accessibility. The organization implements strict access controls
   and enforces Data Usage policies to prevent unauthorized data
   access or misuse. Only authorized individuals or departments can
   access specific datasets based on their roles and responsibilities.
   This controlled approach ensures data protection while still
   allowing appropriate utilization. As the Protectionist culture
   appears in the most sensitive environments, these controls are
   considered enablers of confidence rather than hindrances to
   performance.

3. Strategic Data Partnerships

   Data strategy in a Protectionist culture may involve forming
   partnerships with external entities with complementary data
   assets. This allows the organization to leverage external data
   sources while maintaining control over its own data. These
   partnerships are carefully managed to ensure data security and
   compliance with relevant regulations.

4. Data Analytics for Internal Decision-Making

   While data accessibility is restricted, data strategy in a
   Protectionist culture recognizes the value of utilizing data for
   internal decision-making. The organization establishes internal

data analytics capabilities, where authorized teams are equipped with the necessary tools and skills to perform analysis and derive insights from available data. This enables evidence-based decision-making within the confines of controlled data access.

5.   Continuous Improvement in Data Governance

Data strategy in a Protectionist culture emphasizes the continuous improvement of data governance practices. The organization regularly assesses and updates data governance policies, procedures, and controls to ensure they remain effective and aligned with evolving data protection regulations. This proactive approach enables the organization to adapt to changing data landscape while maintaining a strong focus on data security.

In a Protectionist culture, data strategy aims to strike a balance between control and utilization of data assets. By establishing robust data governance practices, controlling data accessibility, forming strategic data partnerships, leveraging internal analytics capabilities, and continuously improving data governance, organizations can maintain a strong data-driven culture while addressing concerns related to data protection and governance. With a well-executed data strategy, organizations can unlock the value of their data assets while ensuring data security and compliance.

Data Strategy in a Traditional Culture

In a Traditional culture, where data flow is democratized and Data Usage focuses on measurement and diagnostics, organizations rely more heavily on established practices, experience, and intuition in decision-making than the Progressive or Preservationist, where advanced analytics are present. This reliance is generally based on the repetitive nature of the Traditional culture's environment, where multiple measurements may be available, but little inference is used. However, a well-defined data strategy can support the high transaction capability common in the Traditional data culture. In this section, we explore what data strategy looks like within the context of a Traditional culture using the UFDCM.

1. Recognize the Value of All Data

    In a Traditional culture, data strategy begins with recognizing the value of data and its potential to enhance decision-making. The organization acknowledges that data can complement established practices and improve outcomes. A Traditional culture measures extensively and shares data freely in the environment. The organization's supply chain, from marketing to sales to operations, will freely share metrics to ensure orders are processed and fulfilled. Due to the focus on measurement and diagnostics, workers may show disinterest in analyzing processes throughout the supply chain because they are measured on their specific area's performance. Leadership is crucial in driving a mindset of sharing metrics and using those metrics to improve decision-making throughout the supply chain.

2. Augmenting Decision-Making with Timely Data

    Data strategy in a Traditional culture focuses on putting measurement data as close to the decision as possible. The Traditional data culture flows data freely to measure process efficacy or, when anomalies occur, the diagnostic analysis of the issue. The data strategy will endeavor to increase efficacy and responsiveness through tools, training, and infrastructure. With these tools, some predictive analytics and artificial intelligence can be built to enhance measurement and augment decision-making as well.

3. Targeted Data Initiatives

    Data strategy in a Traditional culture involves targeted data initiatives that align with specific business objectives. Rather than a comprehensive data transformation, the organization focuses on identifying key areas where data can have a significant impact. This includes identifying critical metrics and implementing data collection and analysis processes to track performance and inform decision-making in those areas.

4. Supporting Data Literacy

   Data literacy and skill development are necessary in the Traditional culture to support responses to the measurements and diagnostics. The organization provides training programs and resources to enhance employees' understanding of data, statistical analysis, and data interpretation. By fostering data literacy, employees gain the skills necessary to leverage data effectively in their roles.

5. Balancing Tradition and Innovation

   Data strategy in a Traditional culture requires striking a balance between leveraging established practices and the possible disruption from process and measurement changes. The Traditional culture relies on measurement to monitor a process. The data strategy must challenge the measurement, diagnostics, and analytics used in the Traditional culture to increase process innovation. As mentioned, the injection of innovation can be targeted to increase adoption and reduce environmental disruption. This balanced approach ensures that the organization benefits from both the wisdom of experience and the potential of data-driven decision-making.

In a Traditional culture, data strategy must enhance measurement and diagnostic efficacy while introducing innovative thinking. Traditional culture focuses on continuous improvement through measurement with a free flow of data. The organization shares data freely across the organization, focusing on monitoring and improvement. Organizations can optimize the data-driven culture by recognizing the use of all data, augmenting decision-making with timely data, focusing on targeted data initiatives, supporting data literacy, and balancing tradition and innovation. With a well-executed data strategy, organizations can leverage their existing strengths while embracing the power of data to make informed decisions and achieve better outcomes.

Data Strategy in a Progressive Culture

In a Progressive culture, where data flow is democratized and Data Usage includes measurement, diagnostics, understanding, and differentiation, organizations can leverage data strategically to drive innovation, achieve business objectives, and gain a competitive edge. A robust data strategy is critical in supporting and amplifying the data-driven practices and collaborative environment that thrive in such a culture. Let's explore what data strategy looks like within the context of a Progressive culture using the UFDCM.

1. Embracing Data Accessibility and Collaboration

    In a Progressive culture, data strategy focuses on promoting data accessibility and breaking down silos. The organization values transparency and encourages collaboration across teams, enabling data to flow freely. This entails implementing data infrastructure that supports data sharing, employing tools for collaborative analysis, and establishing data governance frameworks that balance security and accessibility.

2. Democratizing Data and Empowering Employees

    A Progressive culture places a strong emphasis on data literacy and skill development for employees at all levels. Data strategy in such a culture involves designing and implementing comprehensive data training programs and resources. These initiatives aim to empower employees to become data literate and capable of using data to make informed decisions and contribute to the organization's success.

3. Cultivating Decision-Making with Data Products

    Data strategy in a Progressive culture must balance the availability of data products against the ability to decide. Data products such as data analytics, data visualization, and advanced analytics techniques allow for uncovering actionable insights. At its best, decision-makers have access to real-time data dashboards and self-service analytics tools, enabling them to make informed, timely decisions based on data-driven insights. These benefits must be balanced against the potential of over-reliance on the data and the spiral of "analysis paralysis." Beyond the technical training, the data strategy must focus on the softer skills of decision-making in the presence of data abundance.

4. Encouraging Proper Data Handling and Security

In a Progressive culture, data strategy fosters a culture of proper data handling and data security. The free flow of data in the Progressive culture increases the potential for inadvertent exposure and external attack. The organization must ensure employees are trained to be able to handle data and share data properly. When internal data sharing is encouraged, data can be shared in insecure ways, such as email. These practices can cause data to be exfiltrated by hackers or accidentally by forwarding. The data strategy will ensure technologies, training, processes, and audits are in place to guard the data assets.

5. Driving Innovation and Future Readiness

Data strategy in a Progressive culture goes beyond operational efficiency; it fuels innovation and ensures future readiness. The organization invests in exploring emerging technologies like artificial intelligence, machine learning, and predictive analytics to uncover new opportunities and drive innovation. Data strategy aligns with the organization's long-term vision, enabling it to adapt and thrive in a rapidly evolving digital landscape.

In a Progressive culture, a robust data strategy serves as a catalyst, amplifying the positive impact of a mature data culture. It lays the foundation for leadership to empower employees, enhance collaboration, encourage experimentation, and position the organization for continuous innovation and growth through data democratization. The data strategy will also ensure the data asset is secure and employees have the training needed to handle the data asset properly. By aligning data strategy with the principles of the UFDCM, organizations can harness the full potential of their data assets and create a sustainable competitive advantage in the data-driven era.

Highly Regulated Industry Data Strategy (A Protectionist Example)

In highly regulated industries such as public utilities, data plays a critical role in shaping operational efficiencies, ensuring compliance with stringent regulations, and making strategic decisions. Within this industry, a Protectionist data culture prevails, characterized by a cautious approach to data management, a focus on data security and privacy, and an emphasis on maintaining strict control over data access and usage. In this context, crafting a robust data strategy that aligns with the principles of a Protectionist data culture is vital to success.

Public utilities face unique challenges related to data governance, risk management, and the complexity of data-driven initiatives. Compliance with regulatory frameworks, safeguarding sensitive customer information, and protecting critical infrastructure assets are paramount. As such, the data strategy must be tailored to cater to these specific needs while fostering a culture that prioritizes data protection, risk mitigation, and adherence to regulatory requirements.

In this section, we delve into the intricacies of developing a data strategy within a Protectionist data culture in public utilities. We explore the key considerations, objectives, and methodologies that must be integrated into the data strategy to effectively navigate the challenges and opportunities in this highly regulated industry. By understanding the core principles of the Protectionist data culture and aligning the data strategy accordingly, public utility organizations can harness data's power while ensuring data security and compliance with regulatory mandates. A data strategy for a highly regulated industry would need to consider the specific requirements and constraints of the industry. Here are some key considerations that a data strategy for a highly regulated industry should include:

1. Compliance: The data strategy should ensure that the company complies with all relevant regulations and laws. This may include data privacy regulations, data retention requirements, and security standards.

2. Data Governance: A robust data governance framework should be established to ensure that data is managed in a consistent and controlled manner. This includes data quality control, data classification, data access controls, and data life cycle management.

3. Risk Management: The data strategy should include a risk management plan that identifies potential risks associated with data management and outlines mitigation strategies. This includes identifying and assessing risks associated with data privacy, security, and compliance.

4. Data Architecture: The data strategy should define the data architecture that is appropriate for the specific industry and the company's needs. This includes the choice of data storage, data processing, and data integration technologies.

5. Data Analytics: The data strategy should include a data analytics plan aligned with the business objectives and regulatory requirements. This includes defining the data analytics tools and techniques that will be used and the data sources that will be analyzed.

6. Data Privacy: The data strategy should include measures to protect the privacy of personal data in compliance with applicable laws and regulations. This includes data anonymization techniques, data masking, and data encryption.

7. Reporting and Auditing: The data strategy should include a plan for reporting and auditing to ensure transparency and accountability. This includes regular reporting on data management practices and auditing of data governance processes to ensure compliance with regulatory requirements.

A data strategy for a highly regulated industry should focus on compliance, data governance, risk management, data architecture, data analytics, data privacy, and reporting and auditing. A well-executed data strategy can help companies in highly regulated industries manage their data effectively, comply with regulations, and gain insights to drive business growth.

Summary

This chapter focuses on the role of a well-designed data strategy in achieving organizational success. A data strategy serves as a compass guiding organizations toward data-driven triumph. Rooted in understanding and embracing organizational culture, the chapter emphasizes the development of a data strategy tailored to each organization's unique cultural context using the Usage and Flow Data Culture Model (UFDCM).

At its core, a data strategy is a blueprint for harnessing, managing, and leveraging data as a strategic asset. It aligns with business objectives, establishes data governance, defines data architecture, supports data management, and fosters a data-driven culture. This comprehensive approach positions data as a catalyst for operational efficiency, customer experiences, innovation, and growth.

Aligning data strategy with organizational strategy yields several advantages. It ensures the right data flows seamlessly to decision-makers, maximizes data value, cultivates a data-driven culture, and drives innovation. The Usage and Flow Data Culture Model (UFDCM) offers a dynamic framework to understand and adapt data strategy, fostering positive cultural shifts that enhance data utilization, decision-making, and competitiveness.

Overcoming challenges in data culture transformation demands addressing cultural resistance, technology adoption, leadership buy-in, talent development, and knowledge sharing. A holistic strategy empowers organizations to transcend data culture barriers, fostering a skilled workforce capable of data-driven decision-making, operational excellence, and innovation. Surveying data culture in different quadrants underscores the need for tailored data strategies. Preservationist cultures prioritize data security, Protectionist cultures focus on controlled access, Traditional cultures balance tradition and innovation, and Progressive cultures empower innovation.

A well-aligned data strategy amplifies organizational success by merging data culture and strategy. This symbiotic fusion empowers decision-makers with the right data, maximizes data value, nurtures a data-driven culture, and sparks innovation. As data strategy and organizational strategy converge, organizations embark on a transformative journey, guided by a data-driven compass toward excellence.

CHAPTER 4

Leadership and Change Management

A compelling story about leadership comes from Stephen Covey's classic book, *The 7 Habits of Highly Effective People*. The story goes like this. "A group of producers are cutting their way through the jungle with machetes. They're the producers, the problem solvers. They're cutting through the undergrowth, clearing it out. The managers are behind them, sharpening their machetes, writing policy and procedure manuals, holding muscle development programs, bringing in improved technologies and setting up working schedules and compensation programs for machete wielders. The leader is the one who climbs the tallest tree, surveys the entire situation, and yells, "You're in the wrong jungle!" But how do the busy, efficient producers and managers often respond? "Shut up! We're making progress."

Leadership plays a central role in shaping the culture of an organization and setting the tone for data-driven decision-making, and that starts with a Chief Data Officer (CDO). Leaders must understand the value of data and champion its usage throughout the organization. They need to communicate the vision, benefits, and strategic importance of building a data culture, ensuring that all employees understand and align with this vision. By demonstrating their commitment to data-driven practices and providing the necessary resources, leaders create a foundation for cultural change.

A CDO must build a coalition of champions and advocates for the data culture. It is essential for driving the adoption of a data culture. These individuals, from different levels and functions within the organization, serve as influencers and change agents. They help promote the importance of data literacy, share success stories, and provide support and guidance to their peers. By fostering a network of data culture ambassadors, organizations can create momentum, encourage collaboration, and overcome resistance to change.

© Gary W. Griffin, David Holcomb 2023
G. W. Griffin and D. Holcomb, *Building a Data Culture*, https://doi.org/10.1007/978-1-4842-9966-1_4

If the CDO is to be successful, then often several data initiatives will be necessary. In short, it's going to require a few changes. Strategies for effective change management, including addressing resistance and fostering a supportive environment, must be considered. Change management is a critical aspect of building a data culture. It involves understanding and addressing resistance to change, providing training and development opportunities, and fostering a supportive environment that encourages experimentation and learning. Effective change management strategies include clear communication, stakeholder engagement, and creating a sense of urgency and shared purpose around the data culture initiative.

Leadership and change management are integral to the successful development of a data culture within an organization. Leaders must take an active role in driving cultural change, building a coalition of advocates, and implementing effective change management strategies. By prioritizing these aspects, organizations can foster a data-driven mindset, encourage collaboration, and create an environment where data becomes an asset in decision-making processes.

Building a data culture within an organization is not simply about implementing new technologies or processes; it requires a fundamental shift in mindset, behaviors, and practices. Leadership and change management play pivotal roles in driving this transformation and creating an environment where data-driven decision-making becomes ingrained in the organization's DNA. In this chapter, we explore the role of leadership and change management in building a data culture and discuss strategies for the successful implementation.

Organizational Structures and the Data Governance Office

Building a data culture within an organization requires a careful examination of the existing organizational structures and their alignment with the desired data culture. As stated in Chapter 2, one of the components of a mature data culture is the creation of a data governance office (DGO). The DGO is a strategic team responsible for measuring success and gathering metrics. The Data Governance Institute defines DGO as "a centralized organizational entity responsible for facilitating and coordinating Data Governance and/or Stewardship efforts for an organization." One excellent way to think about the role of a DGO is that a DGO would be to data what a PMO (Project Management Office) is to projects. This function supports good practices and is the organization's "go to" person for data-related projects.[1]

Why do you need a DGO? Most organizations that begin a formal data governance effort need a support team to facilitate policy creation and enforcement, perform data quality reporting, maintain repositories, and coordinate the activities of councils, stewards, and stakeholders. These traditional data governance functions ensure the Data Flow vector is properly managed for the desired culture (see Chapter 5). Also ensuring flow and facilitating usage as well, the data lake and data warehouse function creates a data repository of operational and strategic data for use in reporting and analytics (see Chapter 7). Similarly, both the usage and flow vectors are affected by the business intelligence, reporting, and dashboarding center of excellence and analytics center of excellence (see Chapter 7).

Who should be included in a DGO? Of course, the answer depends on what you're trying to accomplish with your program. It is our recommendation that you start with a Chief Data Officer to head up the DGO. Most programs include a leader who can navigate politics, organize and facilitate meetings, work with participants behind the scenes, brief executive stakeholders, and make sure research and assignments coming out of council meetings are on track. This person should be a data expert, as they need to be able to converse with all stakeholders and to understand data flow diagrams and data models. Some organizations can't find a DGO leader with all the skills needed. They opt for a leader who can manage politics and people and assign a data architect and/or metadata manager as either full-time members or dotted-line members of the DGO.

Under the DGO, we recommend the CDO will own the traditional data governance functions such as data stewardship, data quality, and security/compliance liaison; process management or liaising with the process management department; data lake and data warehouse function; business intelligence, reporting, and dashboarding center of excellence; and the analytics center of excellence. Other data services in the organization, such as database administration, data architecture, and data modeling, do not fit as well into the DGO or reporting to the CDO; however, some organizations have chosen to have these functions represented in their DGO. These functions require tight integration with the application development and production support organizations. Putting the resources in the DGO does not facilitate better allocation of resources or

---

[1] See the Data Governance Institute website page on establishing a Data Governance Office. It is accessed at https://datagovernance.com/establishing-a-data-governance-office/

adherence to policies. If the DGO includes these organizations, the linkages with IT must be tight for these areas while maintaining separation in others. This separation is necessary to retain objective oversight. For these reasons, we recommend these functions remain in the IT or architecture organizations.

Where should the DGO be placed organizationally? The location and reporting structure of the data governance office (DGO) within an organization can significantly impact its effectiveness and influence on data governance initiatives. There is no correct answer to this question, as the optimal placement of the DGO can vary depending on the organization's size, culture, and strategic goals. Let's discuss the pros and cons of two common placement options for the DGO.

The first option is the DGO is a stand-alone organization with a Chief Data Officer (CDO) reporting to the CEO. There are pros and cons to this type of organizational structure. One pro is that they will have direct access to leadership. Placing the DGO as a stand-alone entity with the CDO reporting directly to the CEO provides direct access to top-level leadership. This ensures that data governance initiatives receive high-level attention and support. A second pro is that it gives the DGO a certain level of independence and authority that may not exist otherwise. The stand-alone position of the DGO may grant it more autonomy and authority to drive data governance strategies across various departments and functions. Third, with direct reporting to the CEO, the CDO can prioritize data-centric approaches without being influenced by other departmental agendas.

The stand-alone organizational structure also has cons associated with it. First, placing the DGO separately may lead to potential silos and hinder integration with other strategic functions within the organization. Second, as a stand-alone unit, the DGO may face resource constraints and budget limitations compared to other established departments. Finally, the CDO reporting directly to the CEO increases the CEO's span of reports, putting pressure on the CEO's already busy schedule.

There are other options, of course. One option that works well is to have the CDO report to a Chief Strategy Officer, and the DGO becomes a part of an established strategy organization. This organizational structure has pros and cons as well. First, it provides a direct line to ensure that the data strategy is aligned with the organizational strategy. This will facilitate the use of data to drive business value, and it will ensure direct accountability to the overall strategic objectives of the company. Second, the DGO can collaborate more effectively with other strategic functions, such as planning, risk management, and business intelligence. Third, being part of a larger strategy organization may lead to better resource allocation and access to budgets for data governance initiatives.

The cons of this structure include a dilution of focus and a direct influence on the DGO's decision-making. Placing the DGO within a larger department might lead to competing priorities and potential dilution of its data governance focus. The DGO's recommendations and decisions may be influenced by the priorities of the strategy organization, which may not always align with the best interests of data governance.

Selecting the correct organization for the CDO to reside in must offer the needed level of autonomy so resources can be focused on data initiatives. For this reason, we recommend the CDO not report to the information technology (IT) organization. The goals of the Chief Information Officer (CIO) and CDO are complementary, but they are not necessarily the same. Resources in the data governance organization will have data skills that are generally technical. These skills can easily be used in IT projects rather than realizing the continual improvement of data quality, accessibility, literacy, and analytics. When reporting into IT, the focus of the CDO tends to be technology focused on IT projects rather than business focused on the usage of data.

The ideal placement of the data governance office depends on the organization's specific needs and strategic priorities. Both options have their merits and challenges. If data governance is considered a top strategic priority and requires strong executive support, the stand-alone model with the CDO reporting directly to the CEO could be more suitable. On the other hand, if the organization aims to align data governance with broader strategic initiatives and ensure collaboration, integrating the DGO within a strategy organization with the CDO reporting to the Chief Strategy Officer might be the better choice. It is crucial for the organization to carefully assess its goals, culture, and existing structures to determine the best fit for the data governance office and its reporting structure. When you're first establishing your data governance office, make sure that your staff have the communication skills and materials they need to work successfully with data stakeholders. Data Governance can be politically tricky.

The Usage and Flow Data Culture Model (UFDCM) provides a framework for understanding the interplay between Data Usage and Data Flow, which in turn influences the organization of work. The DGO will look and function differently within the four quadrants of the UFDCM. This directly impacts the building of a data culture and how the DGO differs in each quadrant.

In a Progressive culture where both Data Usage and Data Flow are high, the organization of work is characterized by collaboration, cross-functional teams, and fluid information sharing. The emphasis is on breaking down silos and promoting a data-driven approach across the organization. Structures such as matrix, team-based, and network structures are commonly found, as they facilitate the integration of diverse perspectives and expertise. The DGO fosters integration of diverse perspectives and expertise, facilitating a data-driven approach across the organization.

In a Traditional culture with high Data Flow but low Data Usage, the organization of work tends to be more hierarchical and functionally oriented. Functional structures are prevalent, with employees working within their specialized domains. Decision-making may rely more on experience and intuition rather than data-driven insights. The DGO can encourage cross-functional collaboration and promote Data Usage in specific projects or initiatives, despite the reliance on experience and intuition in decision-making.

In a Protectionist culture with low Data Flow and low Data Usage, the organization of work is often characterized by centralized decision-making and limited access to data. Hierarchical structures may dominate, and information flows are tightly controlled. The focus is on maintaining control and protecting data assets, which can hinder collaboration and inhibit the development of a data culture. The DGO plays a crucial role in promoting data literacy and transparency to overcome barriers to collaboration and foster a data-driven environment.

In a Preservationist culture with low Data Flow but high Data Usage, the organization of work is characterized by strong governance, privacy measures, and a cautious approach to data accessibility. Structures that emphasize control and protection, such as hybrid or virtual structures, may be present. While access to data may be limited, efforts are made to utilize data strategically for decision-making and maintain the integrity and security of data assets. The DGO ensures that data is strategically utilized for decision-making while maintaining the integrity and security of data assets.

The organizational structures within the four quadrants of the UFDCM significantly influence the organization of work when building a data culture. By understanding the characteristics of each quadrant and aligning the organizational structure with the desired data culture, organizations can create an environment that fosters collaboration, data-driven decision-making, and cross-functional integration. The data governance office (DGO) is an essential element in building a mature data culture. By aligning organizational structures with the UFDCM, the DGO can effectively foster collaboration, data-driven decision-making, and cross-functional integration within each quadrant. With the DGO's support, organizations can establish a robust data culture that optimizes data resources and achieves their mission effectively.

The Role of Leadership in Driving Change and Data Transformation

Effective leadership is instrumental in driving culture change and data transformation within organizations. Leaders have a critical responsibility in setting the tone and direction for data culture initiatives, aligning data initiatives with overall business objectives, and guiding the organization toward becoming data driven (Cortellazzo et al., 2019; Volpato, 2022; Correlation One, 2023; AccelData, n.d.; Overby, 2023; Rice, 2023; Sainger, 2018; Domo, n.d.; Hollister et al., 2021).

In a Progressive data culture, leadership must embody a forward-thinking and innovative mindset. Progressive leaders recognize the strategic value of data-driven decision-making and proactively invest in data analytics capabilities. They champion data culture transformation by promoting a culture of curiosity, experimentation, and learning from data insights. These leaders actively participate in data-driven discussions, use data to support their decisions and strategies, and empower employees to access and analyze data for driving business outcomes (Cortellazzo et al., 2019; Correlation One, 2023).

In a Traditional data culture, leadership must focus on breaking down silos and promoting cross-functional collaboration. Traditional leaders understand the importance of data in decision-making but may face challenges in data accessibility and sharing. They play a crucial role in fostering a culture of data sharing and encouraging departments to collaborate and leverage data collectively. These leaders establish clear data governance frameworks to balance data security and utilization, making data a common language for all business functions (Overby, 2023).

In a Protectionist data culture, leadership must balance data utilization and security. Protectionist leaders prioritize data integrity and safeguarding sensitive information. They implement stringent data protection measures and establish guidelines for controlled data utilization, building trust with employees and customers by prioritizing data security and compliance with regulations (AccelData, n.d.).

In a Preservationist data culture, leadership emphasizes preserving the organization's unique culture and values. Preservationist leaders prioritize continuous improvement analytics and leverage data to stay ahead of internal competitors. They focus on building internal data analytics capabilities and providing the necessary skills and tools to authorized teams for extracting insights from available data sources (Sainger, 2018).

Leadership plays a central role in driving culture change and data transformation in each quadrant of the Usage and Flow Data Culture Model. Whether fostering a data-driven and innovative mindset in a Progressive culture, promoting collaboration and data literacy in a Traditional culture, balancing data utilization and security in a Protectionist culture, or leveraging data insights for continuous improvement in a Preservationist culture, effective leadership sets the foundation for building a successful data culture. Leaders must understand the unique characteristics of each data culture quadrant and tailor their approach to create a data-driven organization that aligns with overall business objectives.

Building a Coalition of Champions and Advocates

To successfully drive a data culture initiative, it is essential to build a coalition of champions and advocates who are passionate about data and can help lead the transformation across different data culture types. The importance of building a coalition of champions can't be overemphasized. Creating a data-driven culture requires support and buy-in from individuals across the organization (Cortellazzo et al., 2019).

Building a coalition of champions and advocates helps in several ways. It provides access to influential voices across the enterprise. Champions and advocates often hold influential positions within the organization, making them effective ambassadors for the data culture initiative. Their endorsement can carry significant weight and influence others to embrace data-driven practices. It can solicit and incite grassroots support. Having a diverse group of advocates from various departments and levels of the organization helps in generating grassroots support for the data culture initiative. This support can create a ripple effect, encouraging others to embrace data-driven decision-making. Advocates can also help break down departmental silos and foster cross-functional collaboration. Their efforts can lead to the sharing of data and knowledge, creating a more cohesive and unified data culture (Cortellazzo et al., 2019; Correlation One, 2023; Overby, 2023).

One of the first tasks is to identify key stakeholders and get their buy-in. Identifying key stakeholders who are enthusiastic about data and data-driven decision-making is crucial for building the coalition of champions. These stakeholders may include data enthusiasts, data innovators, and business leaders. Data enthusiasts are individuals who have demonstrated a keen interest in data, analytics, and its potential benefits for the organization. Data innovators are employees who have previously championed data-driven initiatives or projects, highlighting their willingness to embrace change. Business leaders are executives and managers who recognize the value of data and can use their authority to prioritize data culture transformation (Cortellazzo et al., 2019; Sainger, 2018).

Regardless, there will always be resistance and skepticism, and these must be addressed head on. Resistance and skepticism are common challenges when introducing cultural change. Some stakeholders may be hesitant to embrace data-driven practices due to various reasons, such as fear of losing control, lack of understanding, or concerns about data privacy and security. To address these challenges, transparent and open communication is crucial. Leaders must communicate the vision, benefits, and objectives of the data culture initiative clearly and consistently to address any misconceptions or fears. Providing education and training on data literacy and data-driven decision-making helps build confidence and empowers stakeholders to make informed choices. Implementing small-scale pilot projects allows stakeholders to experience the benefits of data-driven practices firsthand, alleviating skepticism and displaying successful use cases (Volpato, 2022; Correlation One, 2023).

You must carefully implement well thought-out strategies to win over stakeholders. Winning over stakeholders and gaining their support require a tailored approach based on the data culture quadrant. The UFDCM provides guidance on how to address these challenges based on the data culture type. In a Progressive data culture, emphasize the potential for innovation, data-driven insights, and how it aligns with the organization's vision for the future. In a Traditional data culture, highlight how data-driven practices can enhance collaboration, streamline processes, and improve decision-making across departments. In a Protectionist data culture, focus on data security measures and demonstrate how controlled data utilization can lead to better risk management and compliance. In a Preservationist data culture, showcase how data insights can lead to continuous improvement, help preserve the organization's competitive edge, and reinforce its core values.

Building a coalition of champions and advocates is critical to the success of a data culture initiative. Identifying key stakeholders, addressing resistance, and employing tailored strategies to win over support are essential steps in driving cultural change and fostering a data-driven organization. With the right advocates in place, the organization can navigate the challenges and create a sustainable data culture that aligns with its business objectives and flourishes in all data culture types.

Strategies for Effective Change Management

Implementing a successful data culture transformation requires effective change management strategies that address the unique challenges posed by each data culture quadrant. In this section, we examine various strategies for navigating the change management process and fostering a data-driven organization. We discuss how to communicate the vision and benefits of the data culture initiative to employees at all levels. Additionally, we explore effective strategies to overcome resistance to change and address cultural barriers within each data culture quadrant (AccelData, n.d.; Cortellazzo et al., 2019; Correlation One, 2023; Domo, n.d.; Hollister et al., 2021; Overby, 2023; Sainger, 2018; Volpato, 2022).

Effective communication is at the core of successful change management. When introducing a data culture initiative, leaders must clearly articulate the vision, objectives, and benefits to employees at all levels. Strategies for effective communication include tailored messaging, storytelling, and two-way communication. Tailor the messaging to resonate with each data culture quadrant. Highlight how data-driven decision-making aligns with the organization's values and objectives, addressing the unique concerns and priorities of each quadrant. Use compelling stories and real-world examples to illustrate the potential impact of the data culture initiative. This type of approach can help employees connect emotionally with the initiative and understand its significance. Encourage two-way communication, allowing employees to share their thoughts, concerns, and ideas. Actively listen to feedback and incorporate it into the change management approach (Correlation One, 2023; Overby, 2023; Volpato, 2022).

Resistance to change is a common barrier in any cultural transformation, including data culture initiatives. Leaders must address resistance by identifying root causes quickly and understand the reasons behind resistance within each data culture quadrant. Resistance may be driven by fear of the unknown, concerns about job security, or skepticism about the benefits of data-driven practices. Leaders must address employee concerns and misconceptions through transparent communication by providing evidence of successful data-driven initiatives and the positive impact they have had on other organizations. Finally, leaders foster a supportive and empowering environment where employees feel valued and included by encouraging collaboration and participation in decision-making processes.

Leaders must address cultural barriers that can result across the different quadrants of the UFDCM. Different data culture quadrants may exhibit distinct cultural barriers that need to be addressed. Strategies for addressing cultural barriers must be specific to each data culture. In a Progressive data culture, encourage experimentation and risk-taking. Foster a culture that embraces failure as a stepping stone to innovation and continuous improvement. In a Traditional data culture, promote cross-functional collaboration and information sharing. Break down silos and create a culture of data transparency and knowledge exchange. In a Protectionist data culture, emphasize data security and risk management. Build trust in data-driven decision-making by demonstrating the effectiveness of controlled data utilization. In a Preservationist data culture, highlight how data insights can preserve the organization's competitive edge and core values. Show how data-driven practices align with the organization's heritage and long-term objectives.

Effective change management is essential for a successful data culture transformation. By communicating the vision and benefits of the data culture initiative, overcoming resistance to change, and addressing cultural barriers within each data culture quadrant, leaders can foster a culture that embraces data-driven decision-making at all levels of the organization. These strategies empower employees to embrace the change, adapt to the evolving data culture, and contribute to the organization's long-term success as a data-driven entity.

Fostering a Supportive Environment for a Data Culture

Building a supportive environment is a critical aspect of nurturing and sustaining a data culture within an organization. In this section, we explore how leaders can effectively foster a culture that embraces data-driven practices, values data literacy, encourages continuous learning, and promotes open communication. Our focus will be on tailoring these strategies to address the unique characteristics of each quadrant within the Usage and Flow Data Culture Model.

Although this is further discussed in Chapter 6, it's important to note that cultivating data literacy and continuous learning is vital to a mature data culture. In a Progressive data culture, to foster a supportive environment, leaders should focus on providing advanced training. Employees should receive advanced data analytics training, empowering them to leverage data-driven insights for strategic decision-making. Encouraging collaboration and open sharing of data across teams fosters cross-functional learning and innovation, driving the organization forward.

In a Traditional data culture, leaders can foster a supportive environment by offering data interpretation training. Training programs that help employees interpret and apply data findings effectively build confidence in data-driven decision-making. Communicating the reasons behind limited advanced analytics, the focus on measurement and diagnostics, and providing clear guidelines on proper data handling will establish a sense of trust and clarity.

In a Protectionist data culture, to promote open communication, leaders can encourage data discussions. These discussions start with encouraging exposure of data quality issues. Reward systems such as recognition programs or bounties for data quality issues can increase openness.

In a Preservationist data culture, leaders can promote open communication by encouraging data stories and establishing controlled data forums. Supporting teams in sharing data-driven stories that provide valuable insights into internal processes and performance boosts engagement and knowledge sharing. These data stories will demonstrate the use of analytic techniques to realize improvement. This type of sharing can increase the spread of these techniques into other areas for use on the data limited to their domain.

Leaders should regularly recognize and reward data-driven behavior. In a Progressive data culture, leaders can reinforce the data-driven approach by linking performance metrics with data-driven objectives, and recognizing employees who achieve data-backed success drives a culture that values data-driven achievements. Additionally, data champions should be recognized. Identifying and celebrating data champions who actively promote and exemplify data-driven decision-making encourages a sense of ownership and leadership in driving data culture. Highlighting successful data-driven projects and initiatives underscores the impact of data in achieving business outcomes. Introducing data-centric KPIs encourages employees to embrace data in their daily responsibilities and decision-making processes

Leaders play a pivotal role in fostering a supportive environment for a data culture to thrive. By promoting data literacy and continuous learning, encouraging open communication, and recognizing and rewarding data-driven behaviors and outcomes, leaders can create a culture where data-driven decision-making becomes ingrained within the organization, irrespective of its data culture type as described in the UFDCM. A supportive environment empowers employees to leverage data effectively, driving innovation, efficiency, and business growth and enabling the organization to stay competitive and adaptive in a rapidly evolving business landscape.

Summary

This chapter discusses the role of leadership and change management in the process of establishing a data culture. The chapter emphasizes that building a data culture is not just about technology and processes but necessitates a fundamental shift in mindset, behaviors, and practices. The chapter outlines strategies for effective change management and leadership based on the data culture quadrants in the UFDCM and demonstrates how leaders can drive this transformation.

The chapter underscores the pivotal role of leadership, particularly the Chief Data Officer (CDO), in shaping an organization's data culture. Leaders need to comprehend the value of data and advocate its use throughout the organization. Communication of the vision, benefits, and strategic importance of data culture is supreme, accompanied by the provision of resources to establish a strong foundation for cultural change. A mature data culture necessitates the establishment of a data governance office (DGO), responsible for facilitating and coordinating data governance efforts highlighting how the DGO adapts in different data culture quadrants.

The chapter emphasizes the significance of building a coalition of champions and advocates to drive a data culture initiative successfully. Key stakeholders should be identified and engaged, addressing resistance and skepticism. Strategies for each data culture quadrant are presented, showcasing the tailored approaches required to win support. By leveraging influential voices, fostering grassroots support, and addressing concerns transparently, leaders can create momentum and encourage data-driven practices.

Change management is essential for successful data culture transformation. Effective communication, overcoming resistance, and addressing cultural barriers are key elements of change management. The chapter explores strategies for each data culture quadrant, such as tailored messaging, storytelling, and two-way communication.

Leaders play a critical role in fostering a supportive environment for a data culture to flourish. Strategies to cultivate data literacy, continuous learning, open communication, and recognition of data-driven behaviors are outlined. By aligning these strategies with the UFDCM and addressing the unique characteristics of each data culture quadrant, leaders create an environment conducive to embracing data-driven practices. By understanding the nuances of each data culture quadrant, tailoring strategies, and fostering a supportive environment, leaders pave the way for a successful data culture transformation that aligns with the organization's vision and goals.

CHAPTER 5

Data Governance and Infrastructure

Noted Indian politician Piyush Goyal said, "The speed of decision-making is the essence of good governance." Goyal could have easily been talking about data governance because the "speed of decision-making is the essence of good data governance." One might ask, "What about data quality?" Data quality creates trust and increases confidence, allowing decision-makers to decide quicker. Therefore, data quality would be part of "good governance."

Proper data governance enables quick and accurate decision-making by activating the data asset. This activation does not imply more data is available to everyone in the organization. Rather, it means the appropriate amount of data is available to the appropriate people to make the appropriate decision based on their decision rights. Therefore, data governance ensures proper data flow to decision-makers. This flow must be timely, high quality, and relevant.

The word "governance" conjures the ideas of regulation and "bureaucracy." There is no way to soften the reality of oversight of processes, data quality, and information publication. Oversight must occur, so let's consider reducing the friction that's likely to occur. Along the way, we suggest ideas to make these initiatives cultural rather than legalistic. Approaching the ideas from a cultural perspective should make adoption easier and the culture better supported.

From a definition perspective, data governance is a framework of policies, procedures, standards, and guidelines used to manage an organization's data asset creation, maintenance, and use activities. Data governance ensures the flow of accurate, consistent, secure, and accessible data to authorized individuals. Examples of this accuracy and consistency include the efficacy of statistical models, vetting of tools, and ensuring data names make sense.

G. W. Griffin and D. Holcomb, *Building a Data Culture*, https://doi.org/10.1007/978-1-4842-9966-1_5

With accurate and secure data, the users need to be made aware of the availability of the data. The data governance program must include an internal marketing or communication plan for the data. Once users learn of the data available, they will need to understand data meanings, tool usage, and proper data handling. This cycle of new announcements, training, and review is key to the data governance program's success.

A data governance program consists of many initiatives. This book does not present all of the possible initiatives you can implement in a data governance program. Rather, we focus on using data governance to achieve the usage and flow dynamics to realize the target data culture. Data governance will influence many data programs, but they are not part of the data governance program. For example, important programs affecting the usage vector of UFDCM, such as data warehousing, business intelligence, enterprise reporting, and data visualizations, are not data governance programs; however, data governance will influence the storage of personally identifiable information, the process of data sharing, access controls, and other elements. Ideally, data governance and these other important programs would reside in the data governance office and be headed by the Chief Data Officer.

At this point in the process, you have taken several steps. First, you have learned about the UFDCM and how flow, usage, and decision rights work. Second, you have performed an evaluation of some type on your current environment. After this evaluation, you probably saw some areas you want to change. Third, the culture you have is not the culture you want in the future. These three steps prepared you to build your data strategy and the data governance organization and plan.

In this chapter, we cover the setting up of the data governance function in light of the current and targeted (future) data cultures. First, we discuss the interplay between data governance and data culture. Next, we explore the six categories of data governance initiatives. Within each of these categories, we explore several specific initiatives. A cornerstone of a mature data culture is the implementation of a modern data catalog. As cultures have a distinct effect on the focus on the initiatives with each category, we cover each culture type and the nuances of data governance in each. By following the framework, an organization can realize its desired data culture.

Data Culture and Data Governance

The digital age has created significant opportunities for businesses and individuals. First, the amount of data recorded and accessible is orders of magnitude higher than just five years ago. Second, information and communications technology (ICT), including wireless connectivity, Internet infrastructure, telephones and devices, and microstorage, allows the individual to instantly communicate with anyone anywhere at any time of the day. For businesses, these ICT advances have also created new software and service offerings for purchase, creating around-the-clock availability.

With these opportunities also comes the reality of mistakes, abuse, and external bad actors. With so much access and data, unintentional mistakes happen. Disgruntled employees or users may also abuse the access, causing issues. As more companies create new products based on consumers' and businesses' data, bad actors may steal your data. Selecting the proper data culture and implementing solid data governance to support it will reduce the likelihood of internal mistakes or abuse and instill a watchful eye against external bad actors. All cultures will want to safeguard data, but some cultures are more sensitive to security issues than others.

The data governance program must consider the current data culture and the target data culture, especially on the data flow vector. Data flow can have an indirect impact on the usage vector. For example, if data flow is limited, the efficacy and relevance of a statistical model may render it unusable. Additionally, data governance may place compliance restrictions on the usage of data that does flow freely, reducing the opportunity for personalization, etc.

Data governance initiatives typically fall into six categories. The first category is librarian initiatives such as metadata catalogs, business glossaries, report inventories, etc. The second category is compliance and security initiatives, such as compliance implementation, data-sharing agreements, information security interactions, and data retention. The third category is quality initiatives and includes data quality reporting, stewardship programs, change management, and analytic model validation. Fourth, process initiatives include process management, change advisory board (CAB) participation, business continuity, etc. The fifth category includes communication that includes broadcasting updates, release notes for new data elements, and marketing of data availability. The sixth and final category is training. Training includes data training events, involvement in onboard and job transfer changes, and annual compliance training coordination.

The six categories become levers for leadership to pull when moving the organization right (more democratization) or left (more controls) on the flow vector. For example, an organization currently in the Progressive quadrant has encountered an event such as an inadvertent exposure or an employee downloading all their accounts before leaving the organization. The event could be a new contract in a governmental marketplace where data sharing is restricted. In either case, leadership has decided to move to the left on the data flow vector to be more controlled. This movement can be a subtle one that may not change quadrants but move the organization closer to the leftmost area of Progressive culture but not quite in the Preservationist quadrant.

Each category has a unique array of tools, roles, responsibilities, and outcomes. Some have organizations, meetings, and forums specific to data governance. In others, tools and technology underpin the category's deliverables. Other categories rely heavily on interactions with other departments such as legal, compliance, human resources, or IT. Data users will be producers, consumers, or both in some cases. In the case of data stewards, the individuals are frequently "voluntold" by their boss and congratulated on their inclusion in the initiative. As we cover the categories and the involvement needed, certain principles should guide the "selling" of the process.

The usage vector does play a role in the implementation of these initiatives. As usage increases to more advanced exploitation of the data, an increase in the flow of metadata, compliance requirements, training, etc., will naturally follow. For example, when data is used only for measurements, data governance efforts for statistical model validation, data quality measurements, or any librarian initiatives will be reduced. Generally, fewer data elements are needed, which may cause a reduction in the specific initiative's scope. Precautions should be in place to not restrict flow due to undersized infrastructure to handle the additional information. Leadership could decide to move up the usage vector, causing increased data flow and a need for more data governance services.

The movement across the data flow vector means organizations will move between the Protectionist and Traditional cultures or the Preservationist and Progressive cultures. Suppose leadership decides to limit the amount of data in a Progressive culture, as described earlier. In that case, they will become a Preservationist culture unless a separate decision is made to change their usage behavior. This movement across the data flow vector must be managed carefully. Attempts to quickly pull back access can be received negatively; if done too slowly, the results will be deferred. The authors' experience shows if done too slowly, the user will find a workaround to continue the flow. Likewise, increasing data flow too quickly can cause confusion and mistakes in decision-making. Leadership must regulate the flow changes purposefully and monitor them closely.

These movements across the data flow vector should achieve the target data culture when coupled with the Data Usage vector decisions. The pace for pulling the levers of data flow via the six categories of data governance initiatives is much like driving your car on a dirt road. If the ride is going smoothly, you may speed up to reach the destination more quickly. There is a case to be made to keep the cadence. The smoothness may be because you have found cultural tolerance for change. If the ride is bumpy, you may want to slow down. There is a case for keeping the speed even with the bumpiness, as going slower has real effects and costs. Next, we explore the six categories of data governance initiatives, their impact on data flow, the indirect effect on the usage vector, and the cultural effects during the movement.

Data Governance: Librarian Initiatives

Librarian initiatives encompass a suite of practices and products focused on inventorying the organization's data assets and products. These initiatives include data documentation, lineage, and stewardship. At its core, this category serves as a key facilitator of data understanding by creating infrastructures such as metadata catalogs, business glossaries, and report inventories. In addition to the infrastructure, the librarian initiatives create the organizational structure to ensure the data descriptions and lineage are high quality. This organizational component consists of data stewards throughout the organization with domain expertise.

The entire organization benefits from the products of librarian initiatives called librarian services. For leadership, librarian services can increase the return on investment of data projects, identify over- and under-supply information flow, and instill data ownership. For technology personnel, data lineage can be evaluated to determine cascading effects of changes made to technologies. Report builders can leverage definitions and lists of values. Data product consumers in the functional organization benefit by having inventories of products for selection. Librarian initiatives should be pursued to optimize the data asset regardless of the current or targeted culture.

Librarian services can be used to move left and right on the data flow vector. Having the data on the inventory of data products can allow leaders to manage data supply or flow. For example, an analysis of the organizational design compared to the report inventory may show high consumption in one area and very little to none in another. Leadership may find parts of the organization have almost no report consumption. Consider in this scenario that the leadership would like to increase data flow. Leaders could evaluate those areas of heavy consumption to replicate the behavior in those lower-consuming areas. Conversely, leadership may see data used by an organization that should not have access and determine whether a reduction of data flow should occur. Having the inventories offers leadership tools to ensure the data assets are optimally used.

Modern Data Catalog (Cornerstone of Librarian Services)

Within the librarian initiatives, the data catalog is the gold standard. The catalog offers knowledge of the data assets within the organization, reducing costs for data integration, new system implementation, and process changes. This section explores how a modern data catalog supports the implementation of enterprise data platforms, enterprise business intelligence, data quality management, master data management, and performance management, ultimately fostering a mature data culture.

1. Centralized Metadata Management

    A modern data catalog is a centralized repository for storing and managing metadata, including definitions, lineage, sources, and usage information. By providing a complete view of the organization's data assets, the data catalog enables effective discovery and understanding of available data resources. Centralized metadata management supports the implementation of various technologies by providing a comprehensive and up-to-date reference for data consumers and practitioners.

2. Metrics and Business Terms

    In addition to the data elements in the central repository, metrics definitions and business terms are also stored. This combination of data elements and metrics allows the practitioner to analyze current business rules and evaluate systems to determine potential effects. This line of sight of data from systems to metrics and business terms will also aid in the training of new employees.

3. Data Product Inventories

    Organizations should maintain a central repository of data products such as dashboards, analytics, or reports. The data products are created by the organization from the data in systems. The inventory repository includes these data products like reports, dashboards, and visualizations, along with their access controls. When coupled with the data, metrics, and business terms repository, the report inventory creates a complete supply chain of data from origin through reporting.

4. Advanced Functions of a Modern Data Catalog

A modern data catalog supports data governance efforts by enabling organizations to establish data policies, standards, and guidelines within the catalog itself. It allows data stewards to define data quality rules, access controls, and Data Usage policies, ensuring adherence to regulatory requirements and organizational best practices. The data catalog also promotes collaboration among data stakeholders, fostering a culture of transparency, accountability, and collective responsibility for data management.

Once implemented, a modern data catalog can increase the effectiveness of integrations, allow better control of the data assets, and enhance collaboration. Just as the organization's data is used for reporting and process improvement, the data catalog has many uses to optimize the data in the organization.

1. Data Asset Discovery and Accessibility

A modern data catalog facilitates data asset discovery by offering search and browsing capabilities. It allows users to explore available datasets, understand their structure, and access relevant data sources. By providing a user-friendly interface and comprehensive metadata, the data catalog improves data accessibility and promotes self-service analytics. This accessibility enables data consumers to leverage enterprise data platforms, enterprise business intelligence, and other technologies effectively.

2. Integration with Data Quality and Master Data Management

By integrating with data quality management and master data management solutions, a modern data catalog enhances data reliability and consistency. It provides visibility into data quality metrics, lineage, and validation rules, enabling data consumers to assess the quality and fitness for use of data assets. The catalog's integration with master data management allows users to access authoritative and consistent master data, supporting accurate analysis and decision-making.

3. Performance Metrics and Data Usage Insights

    A modern data catalog can capture and report performance
    metrics related to data assets, such as Data Usage, popularity,
    and user feedback. These metrics provide valuable insights into
    the effectiveness and impact of implemented technologies like
    enterprise data platforms and enterprise business intelligence.
    By monitoring and analyzing Data Usage patterns, organizations
    can optimize their data infrastructure, improve data accessibility,
    and align their technology investments with user needs and
    preferences.

Data Governance: Compliance/Security

The data governance organization will quickly become friends with the compliance,
legal, information security, records management, privacy, human resources, and
information technology (IT) organizations to ensure the compliance and security needs
of the organization. Each organization plays a unique role in ensuring the organization's
data is used and secured properly. Some organizations may not have all the departments
mentioned previously; however, the functions exist to ensure the organization
complies with regulations, industry standards, and laws for the jurisdictions where
the organization resides or does business. The data governance organization must
collaborate and liaise with these organizations for success.

1. Compliance Organization: In many companies, the compliance
    function is shared among the subject matter experts and their
    associated departments. For example, finance handles financial
    compliance, especially for reporting results to the street. Data
    governance would access the requirements, such as rules,
    approvals, and cadence, to ensure proper handling of data to meet
    compliance.

2. Legal Organization: Data governance will work closely with legal
    when audits, data retention, etc., are needed. Data governance
    must know these exceptional events to ensure data are properly
    executed. Legal is a key contributor to creating policies and
    procedures for data handling and disposal.

3. Information Security: Data governance plays a key role in cyber security efforts for an organization. The information security organization will be on the front lines of data protection, and the data governance organization must coordinate. Publishing data-handling policies, privacy policies, data collection disclaimers, and data classification policies is an area of collaboration between data governance and information security.

4. Records Management: Records management defines what is a "corporate record." This delineation is legal and must be understood to govern the underlying content of the data. Additionally, the data associated with company records have distinct rules for retention and use. These rules may require special handling.

5. Privacy Department: Depending on the organization, the level of privacy may be raised to its own organization. The privacy organization deals with special situations of personally identifiable information (PII) where the specific privacy of individuals must be maintained during the process—for example, cryptocurrency exchanges, anonymous money transfers, medical registries, etc.

6. Human Resources: Human resources is unique for data governance. Frequently, the data in human resources carries higher confidentiality than other business data. This data includes personnel records, reviews, benefits selected, salaries, sex, age, etc., and requires special handling.

7. Information Technology (IT): Most data is held within the systems managed by the information technology (IT) organization. For this reason, IT will be an important partner in data governance and compliance. Many cases exist where data governance must inform IT of new developments that require technological changes such as encryption, obfuscation, or masking to meet compliance standards. Additionally, the IT organization has practices such as test data creation, application upgrades,

integrations, and migrations requiring specific data policies and audits. Finally, IT will interface with external organizations for APIs, data transfers, etc. These organizations must attest to data privacy and security for compliance.

The data governance organization will interact with all departments, but these departments will be integral to building policies, practices, and promotions. Creating an interaction model where these groups periodically meet to discuss policy and communication will facilitate reaching the target data culture. These groups influence Data Flow with their controls and procedures. Data governance must advocate for policies and controls that support the desired flow level and, ultimately, target culture.

Data Governance: Quality

Data quality initiatives ensure the integrity of data. As data enters the organization through input screens, purchased datasets, and transactions; is created by data manipulation processes such as analytics; or is modified or deleted by processes, gates must be established to ensure data is accurate. Mature data governance implements processes and technology such as master data management, data stewardship, data quality reporting, and data entry requirements.

Data stewards play an important role in data quality. Earlier, data stewards were mentioned as subject matter experts populating the modern data catalog. In that population, the data stewards update lists of values, data ranges, etc. Data stewards work within functional areas of the organization, such as marketing, finance, and sales. If data comes into the environment in their domain of expertise, the data steward would be consulted to bring the data into conformity.

Master data management (MDM) can increase the efficacy of data and data stewardship activities. MDM focuses on establishing and maintaining a single, trusted view of critical data entities across the organization, such as customers, products, or locations. MDM platforms enable organizations to consolidate, standardize, and synchronize master data from various sources, ensuring consistency and accuracy. Data stewards can arbitrate disputes and ensure data is shared appropriately when data conflicts across the systems. By establishing a reliable foundation of master data, organizations can enhance data integration, enable cross-functional collaboration, and support accurate and reliable reporting, fostering a data culture that values data consistency and a single version of the truth.

Data governance will actively set guidelines and standards for data entry and quality protocols. These processes will weigh the cost differences of time to enter versus data quality. For example, suppose a screen asks for the state for an address. The field could be edited using a drop-down menu only allowing specific values. If a new value exists, the edit list must be updated. Until the value is in the drop-down, it cannot be selected. Likewise, setting the value from the drop-down can take the user's time. In this case, the potential for abandonment may be too costly. Alternatives could be offered that speed input but may adversely affect data quality. Data governance would memorialize the decisions and ensure processes are implemented to support the decisions.

Not all data can be managed with edits or master data management techniques. Data governance will implement data quality reporting showing the conformity to current known standards or changes in the distribution of values in a field. Data stewards and functional personnel can correct the anomalous values from these reports. Realizing these corrections may require updating multiple systems; the data governance would develop an update process in conjunction with the data stewards and systems owners. Measuring data quality must have a remediation component to be effective.

Data Governance: Process Initiatives

Data is created in business processes. In a keynote address for a business intelligence symposium, co-author Dr. David Holcomb (2010) stated, "Data is the artifact of a process. It does not spontaneously generate or spontaneously combust." This quote makes the case that data will not exist unless something in the ecosystem occurs. Sales order data comes from sales processes; accounting entry data comes from the accounting process; marketing program data comes from the marketing process. For this reason, business process management must be part of good data governance.

Data governance must attack business process management from two perspectives. First, data governance must inject itself into existing processes. These processes include program management office and project life cycle, change advisory boards (CAB), vendor management, and systems procurement. During these existing processes, data governance must insert gates to ensure the proper consideration for data confidentiality, integrity, and accessibility. Second, data governance must plug into the organization's process management if the data governance office does not own the practice. Process mapping and management offer tremendous value to the organization on many fronts, with data integrity among them.

Data governance must consider the organization's current business processes. Established processes such as project approval, change advisory board (CAB), or vendor selection should include data governance artifacts. For example, CAB should have a requirement to update metadata and master data management for any data change and updates to all process maps representing any change. Data governance should ensure vendor management includes compliance attestation, agreement with data-sharing protocols, and data processing agreements.

Business process management should be a focus of data governance as well. Process management is the discipline of documenting processes using flow chart techniques to capture the inputs, processes, and outputs. These flow charts capture the suppliers of inputs, consumers of the outputs, and any dependencies on other processes. Suppliers and consumers can be individuals, groups, external organizations, or other processes. An important aspect of process mapping is capturing the data definitions from the input data, any data used in the process, and the output data. The process mapping will also capture any rules in the process.

Process maps offer tremendous value for improvement and change. When process changes are undertaken in the form of new systems, changes in steps, or new steps, the organization can begin with the current process map and associated data definitions. This starting point increases the pace of change, reduces discovery costs, and enhances the likelihood of success. Finally, process maps effectively educate new employees or vendors on the processes, increasing onboarding speed.

When considering the flow vector of the UFDCM, business process management can give leaders tremendous insight. With business process management, leaders know the suppliers, consumers, and data in the organization's processes. This knowledge can be used to limit data access to only those individuals within the process. Alternatively, the data could be democratized to those involved in the downstream, upstream, or both processes. Data governance and process management offer the maximum flexibility in moving across the data flow vector.

Data Governance: Communication

Data governance must promote policies, practices, changes to data, data products, and the flow of data governance products. Data governance policies require communication to the masses for initiation and changes. Promoting the data catalog and data changed by releases requires focus and an understanding of the users. Finally, data governance includes several products that should be marketed throughout the organization, such as report inventories. By establishing a communication infrastructure, the messages from the data governance will have a higher likelihood of success.

Communicating policies and practices is vital for organizational success, but it can be dry. When crafting policies, the data governance organization should reference the style utilized for other corporate policy documents. In addition to formatting based on existing policy documents, selecting the proper method to communicate changes is important for success. When deciding how to share this type of information, consider the organization's cultural norms for tone and communication cadence to avoid fatigue. Coordination with other policy announcements and creating recurring communications, such as a newsletter, can produce positive results.

Data governance has many products to market. A mature data governance environment includes the mature data catalog, which contains a report inventory, data descriptions, and metrics definitions. These products offer the data governance organization marketing opportunities. These products allow the data governance organization to interact with the entire user base with a very positive and supportive message.

The communication should include branding of the data governance office to get the word out. Communications plans should consist of events such as data summits, lunch and learns, and newsletters. The data governance organization will have meetings to support activities such as data stewardship and leadership updates. Using other forums, such as attending staff meetings or presenting during larger corporate gatherings, can give data governance a positive image.

Data Governance: Training

Procedures, tool literacy, data meanings, and governance processes must be part of the training program to ensure employees understand all of these areas. Throughout these initiatives, policies and procedures have been created. Adoption is dependent on education and monitoring. Data meanings and training on the tools in the data governance suite will increase the effectiveness and efficiency of the organization using the data. Data stewards require training on the processes and tools to enhance data quality. Finally, coordination with other organizations should ensure a cohesive message is presented.

At its core, data governance is governing. Policies and procedures must be implemented, monitored, and trained. When these policies and procedures limit data flow, education must help the employee understand the necessity of the limitation. Crafting the correct tone and message is necessary. Not all messages will be easy, but resistance can be reduced with the right tone, cultural considerations, and outcomes focus.

Data governance is also business value creation through education on the data asset. When an organization invests in technology, people, and time to create a modern data catalog, training on the data meanings can yield significant value. This training will increase technical and functional individuals' Data Usage effectiveness. The culture's usage will be more precise when these individuals become more effective. For example, outcomes may include measurements that will increase focus, analytics that will discover insights faster, and diagnostics that will lead to quicker implementation of remediations and preventive measures. In Chapter 6, we discuss data literacy further.

Education must be accompanied by measurement. Policy adoption can be measured by exceptions or the number of cycles of conformity. Frequently, we hear of policy violations; however, measures can be focused on the absence of failures. Assembly lines, shop floors, and even highways tout the number of days since the last incident of nonconformity. Postings such as "100 days since the last incident" have three important facets: (1) it's an achievement, (2) employees know things are measured, and (3) it sets a standard for future performance or to continue the streak.

Integral to data governance is well-trained data stewards. The tooling in the modern data catalog can facilitate the collection and assimilation of knowledge about the data assets. This knowledge must include the business context of the data. The data stewards possess this knowledge. Training them on using tools and processes allows that knowledge to be collected and shared. Remember, the data stewards have "day jobs" within their functions, so training efforts need to be considerate of their time. This time element is a delicate balancing act that data governance leadership must recognize and work with functional leadership to provide data stewards support.

Finally, training coordination with other organizations is vital. When an employee is hired or changes positions, data governance will be involved. The employee needs access to the proper data and to be trained on its meaning. Similarly, when software is introduced or upgraded, data governance will be involved to ensure training and data definition support. The liaising activities with organizations for security/compliance, processes, and communication mentioned earlier must also include training.

Data Governance in Each Culture Type

As previously mentioned, data governance has a direct impact on the data flow vector and an indirect impact on the Data Usage vector. In general, the limitation of data flow can reduce the users' freedom on the Data Usage vector. Similarly, a reduction in the Data Usage vector allows for a reduction in the diversity of the data governance required. As mentioned earlier, if the organization does not have statistical models, then no librarian, compliance/security, quality, communication, or training services on statistical models are necessary. Data governance services must be considered in each category referencing statistical models if the organization adds statistical models. Understanding each culture's nuances will help when setting up the data governance function and responding to inquiries or migrating between cultures.

Data Governance in a Progressive Culture

In the Progressive culture, data governance embodies innovation, adaptability, collaboration, and empowerment, fueling a dynamic exploration of data's transformative potential. Data governance embraces emerging technologies and encourages cross-functional cooperation. Data flows freely, fostering creativity and experimentation, while agile governance structures ensure responsiveness to change.

With the Progressive culture's perspective in mind, the following outlines some distinctive features driving each category of initiatives:

- Librarian Services: These initiatives must focus on constant change and collaboration. Adopting advanced analytics requires librarian services to contain more metadata and data asset descriptions than measurement or diagnostic-only cultures.

- Security/compliance initiatives embrace innovation while managing risks. The focus is on fostering a culture of transparency and accountability through open communication, continuous monitoring, and rapid incident response strategies.

- Process initiatives are guided by innovation and agility in data-related processes. Data governance partners with cross-functional teams, technology experts, and process innovators to implement flexible and adaptive workflows focused on leveraging technology, automation, and data-driven insights to optimize processes and drive continuous improvement.

- Quality initiatives focus on the needs of a comprehensive dataset, content validity, and derivative data analytic products. With the full spectrum of analytics, data quality will include capabilities for model validation. As in all cultures, data quality is key to effective decision-making.

- Communication initiatives will be similar in all cultures; however, Data Usage and technology adoption levels can offer opportunities and challenges. The availability and adoption of personalization and advanced analytics provide unique opportunities to communicate using electronic means with targeted messaging. The highly collaborative nature of Data Usage offers opportunities for communication about data as long as those messages do not include data restrictions that run counter to the Progressive culture.

- Training initiatives will be in high demand as the usage of the data products will be increased. The focus will be on self-service and enablement to allow users to access knowledge about data easily and quickly.

Data Governance in a Preservationist Culture

In the Preservationist culture, data governance is characterized by a thorough, methodical, patient, and focused commitment to preserving data's essence and context for enduring significance. Data stewardship takes center stage here, meticulously curating data to maintain historical accuracy and cultural relevance. Governance initiatives prioritize data lineage, comprehensive documentation, and meticulous archiving, ensuring data's longevity. The Preservationist mindset requires data governance to focus on the tradition, culture, and heritage within the data asset.

With the Preservationist culture's perspective in mind, the following outlines some distinctive features driving each category of initiatives:

- Librarian services are regarded as custodians of data heritage and legacy. The modern data catalog is meticulously curated, capturing the essence of historical data artifacts, lineage, and business context. The library services may be limited to selected datasets based on priority and need-to-know.

- Security/compliance initiatives are guided by the principles of historical accuracy and data preservation. Data governance collaborates with legal, privacy, and records management departments to ensure that data is protected and retained in accordance with established guidelines. The focus is on protecting and preserving data assets as a reflection of the culture's reverence for tradition and historical significance.

- Process initiatives center around maintaining historical continuity and authenticity in data-related activities. Data governance collaborates with archival experts and process custodians to document and safeguard traditional processes while incorporating modern technology where appropriate. The focus is ensuring data-related processes reflect historical accuracy and authenticity while adapting to modern requirements.

- Quality initiatives focus on the needs of a limited dataset, content validity, and derivative data analytic products. With the full spectrum of analytics, data quality will include capabilities for model validation. As in all cultures, data quality is key to effective decision-making.

- Communication initiatives will be similar in all cultures; however, Data Usage and technology adoption levels can offer opportunities and challenges. With data limitations inherent in the culture, the communication will be focused on the datasets available, their data meanings, and security, compliance, privacy, and sharing limitations. The availability of a full range of analytics may lead to some personalization focusing on operational needs.

- Training initiatives will be in high demand as data product usage increases; however, the data accessible to the pupil will be limited based on leadership decisions. In the Preservationist culture, coordination with the information security, compliance, and records management organizations will yield great benefits for consistent messaging.

Data Governance in a Traditional Culture

The Traditional quadrant prioritizes a steadfast, structured, measured, conservative data governance approach grounded in stability and established practices. Data governance is characterized by well-defined processes, thorough documentation, and adherence to proven methodologies. Data is treated as a precious asset, managed meticulously to ensure accuracy and consistency. Governance frameworks are time tested and risk mitigation focused, delivering reliability and predictability.

With the Traditional culture's perspective in mind, the following outlines some distinctive features driving each category of initiatives:

- Librarian services embody stability and reliability and serve as a reference guide, reflecting the organization's enduring values and practices. The modern data catalog is a well-organized repository of established data assets, documenting time-tested definitions, lineage, and report inventories. Data stewards, often respected veterans in the organization, ensure that data descriptions reflect the tried-and-true conventions.

- Security/compliance initiatives center around upholding established norms and practices while ensuring data security. Data governance works closely with compliance, legal, and IT departments to ensure data handling aligns with industry regulations and internal policies. Maintaining stable data management processes, conducting regular compliance audits, and safeguarding data assets through controlled access and monitoring are the focus.

- Process initiatives prioritize the preservation of established workflows and practices. Data governance works closely with functional teams and process owners to define standardized data-related processes and procedures, ensuring consistency and protocol adherence and minimizing process variations.

- Quality initiatives focus on a comprehensive data selection and its content validity. With data analytics focused on measurement, diagnostics, and some predictive capabilities, quality initiatives will be focused on operational improvement with a limited innovation focus. As in all cultures, data quality is key to effective decision-making.

- Communication initiatives will be similar in all cultures; however, Data Usage and technology adoption levels can offer opportunities and challenges. With no data limitations, the communication may focus on current data needs and the expansion of data that could be used for measurement. Since the Traditional culture focuses on measurement and diagnostics, the communication will focus on the available datasets, their data meanings, security, compliance, privacy, and sharing limitations.

- Training initiatives emphasize continuity and respect for established practices. The focus is on transmitting knowledge with a focus on the operational and tactical methods common in a measurement and diagnostics-focused culture. Although data is not explicitly limited or restricted, the focus on measurement and diagnostics causes training to focus on the data associated with the job.

Data Governance in a Protectionist Culture

Embracing a guarded, controlled, selective, and risk-averse stance, the Protectionist quadrant safeguards data integrity and confidentiality through vigilant data governance. Data protection is paramount, with stringent access controls, encryption, and compliance measures in place. Governance processes prioritize security audits, regular assessments, and proactive threat detection. Data is closely monitored and tightly controlled, minimizing exposure to potential breaches.

With the Protectionist culture's perspective in mind, the following outlines some distinctive features driving each category of initiatives:

- Librarian services serve as gatekeepers, meticulously controlling data access and dissemination. The modern data catalog is a controlled repository, carefully managing metadata, classifications, and access privileges. The library services may be limited to selected datasets based on priority and need-to-know with an eye toward security and privacy.

- Security/compliance initiatives focus on creating a fortified data environment. Stricter access controls, encryption protocols, and data segregation are implemented to prevent unauthorized access.

- Process initiatives emphasize meticulous control and oversight of data-related processes. Data governance collaborates with process owners, compliance teams, and change advisory boards (CAB) to establish rigorous change management protocols. The focus is minimizing risks and disruptions to data flows by implementing well-documented processes, thorough impact assessments, and controlled change procedures.

- Quality initiatives focus on the needed data and its content validity. With usage focused on measurement, diagnostics, and little else, quality initiatives will be focused on operational use. As in all cultures, data quality is key to effective decision-making.

- Communication initiatives will be similar in all cultures; however, Data Usage and technology adoption levels can offer opportunities and challenges. With data limitations inherent in the culture, the communication will be focused on the datasets available, their data meanings, and security, compliance, privacy, and sharing limitations. The Protectionist culture values measurement and has limited analytics usage, requiring focused communications pointed at the "need-to-know."

- Training initiatives will be focused on the operations and data available. The data accessible to the pupil will be limited based on leadership decisions. Likewise, the ability to analyze it with advanced analytics is restricted. In the Protectionist culture, coordination with the information security, compliance, and records management organizations will yield great benefits for consistent messaging.

Summary

This chapter offers the practitioner a framework for establishing a data governance function. First, the relationship between data governance and data culture is explored. Six categories of data governance initiatives are outlined as part of the exploration, including librarian, quality, process, security/compliance, communication, and training. These initiatives look different based on the culture you currently have and want to build for the future. One specific initiative, the modern data catalog, is a centerpiece to increasing the adoption of data governance and the targeted culture.

As leadership considers the current environment, programs currently in place, and the movement to the new culture, relationship building is a great place to start. You cannot go wrong by building the relationships mentioned in the security and compliance initiatives section. These organizations have a vested interest in data governance success because it supports their initiatives. These relationships offer reciprocal benefits.

While you are building these relationships, look back to the assessment of the current environment and leverage it for the next steps. When you assessed the current culture, you also made decisions regarding the target culture. With the earlier assessment, your target culture selected, and information in this chapter, you can propose data governance initiatives to begin the movement. These decisions require technology, people, process, time investments, and ultimately funding. A well-crafted data governance program proposal will include true business value through high-quality data with appropriate accessibility.

CHAPTER 6

Data Literacy and Skills Development

There's an old Chinese proverb that says, "When planning for a year, plant corn. When planning for a decade, plant trees. When planning for life, train and educate people." In today's data-driven world, organizations across industries are recognizing the immense value that data holds. To fully leverage this potential, building a data culture has become a strategic imperative. At the heart of a data culture lies data literacy, a fundamental competency that empowers individuals to effectively navigate, interpret, and communicate data to drive informed decision-making. This chapter explores the concept of data literacy within the context of a data culture, highlighting its significance and implications for organizations aiming to harness the power of data.

Data literacy can be defined as the ability to understand, evaluate, and work with data effectively. It encompasses a range of skills, including the capacity to interpret data, ask relevant questions, analyze information, and derive actionable insights. Data literacy is not limited to technical proficiency with data tools and technologies; rather, it encompasses a broader understanding of data's context, limitations, and ethical implications.

In the context of a data culture, data literacy becomes a foundational skill set that empowers individuals at all levels of an organization to engage with data confidently. It enables employees to become critical consumers and producers of data, fostering a data-driven mindset that permeates decision-making processes and drives innovation. Data literacy is crucial for several reasons within a data culture:

1. Informed Decision-Making: Data-literate individuals possess the skills to access, analyze, and interpret data, enabling them to make informed decisions based on evidence and insights rather than intuition or guesswork. This leads to better business outcomes and mitigates the risks associated with data-driven decision-making.

© Gary W. Griffin, David Holcomb 2023
G. W. Griffin and D. Holcomb, *Building a Data Culture*, https://doi.org/10.1007/978-1-4842-9966-1_6

2. Data-Driven Collaboration: A data culture promotes collaboration and encourages cross-functional teams to work together on data-driven projects. Data literacy allows individuals to effectively communicate and share insights, facilitating collaboration between data experts and domain specialists.

3. Trust in Data: Data literacy promotes a healthy skepticism toward data, encouraging individuals to question data sources, validity, and potential biases. It equips employees with the ability to identify misleading or inaccurate information, fostering a culture of data integrity and trust.

4. Democratization of Data: By building data literacy across an organization, data becomes accessible and understandable to a wider audience. This empowers employees across various roles and departments to explore and analyze data, driving innovation and allowing for diverse perspectives in decision-making.

5. Ethical Use of Data: Data literacy includes an understanding of ethical considerations related to data collection, usage, and privacy. In a data culture, data-literate individuals are better equipped to manage sensitive data responsibly, ensuring compliance with regulations and ethical guidelines.

6. Data Handling: Procedures for handling, transmitting, and storing data while in the user's possession. For example, if the policy states the user will transfer data with a secure password, training on data handling would teach the procedure.

Promoting Data Literacy in a Data Culture

Creating a data culture requires intentional efforts to foster data literacy throughout an organization. It involves a combination of educational initiatives, organizational support, and a shift in mindset. There are several key strategies that can be useful in promoting data literacy in a data culture.

- First, provide comprehensive data literacy training programs that cater to employees at all levels of technical expertise. These programs should focus not only on technical skills but also on critical thinking, data interpretation, and effective communication of insights.

- Second, identify and empower individuals within the organization who exhibit strong data literacy skills to function as advocates and mentors. These champions can help bridge the gap between technical experts and nontechnical employees, fostering a collaborative environment.

- Third, emphasize the importance of data visualization and storytelling techniques to effectively communicate data insights. Encourage the use of visualizations and narratives to present data in a compelling and understandable manner.

- Fourth, invest in user-friendly data tools and technologies that enable employees to access, explore, and analyze data without significant technical barriers. Implement data governance practices that ensure data quality, security, and privacy.

- Fifth, establish a culture of continuous learning and feedback by providing ongoing opportunities for employees to enhance their data literacy skills. Encourage employees to apply their data skills to real-world projects and providing feedback to foster growth and improvement.

Data literacy forms the foundation of a data culture, enabling individuals to engage with data confidently, make informed decisions, and drive organizational success. In a data-driven world, organizations must prioritize data literacy initiatives to empower employees at all levels with the skills to effectively navigate and utilize data. By fostering data literacy within a data culture, organizations can unlock the full potential of data, foster collaboration, and drive innovation.

Designing and Implementing Data Training Programs

Data literacy is a critical skill for individuals and organizations in today's data-driven organization. To effectively harness the power of data, organizations must design and implement comprehensive data training programs and resources. These initiatives play a vital role in fostering a data-literate workforce and promoting a culture of data-driven decision-making. In this section, we explore the key considerations and strategies for designing and implementing data training programs and resources tailored to different data culture types based on the UFDCM.

Before designing data training programs, it is essential to recognize the significance of data literacy in each data culture type. As an example, Gummer and Mandinach (2015) write that data literacy for teachers involves multiple domains, including disciplinary content knowledge, pedagogical content knowledge, and data use for teaching knowledge and skills. In a Progressive data culture, where Data Usage is advanced and data flow is democratized, training should focus on advanced data analytics skills and promoting a collaborative data-sharing environment.

In contrast, a Traditional data culture emphasizes measurement in Data Usage with free data flow. For this data culture type, data interpretation training and clear guidelines for Data Usage within approved boundaries are crucial (Gummer & Mandinach, 2015). For a Protectionist data culture, where Data Usage is basic and data flow is limited, promoting open communication and data discussions is essential. This data culture type requires employees to feel comfortable sharing insights and concerns while working with limited data access. In a Preservationist data culture, Data Usage is advanced, but data flow is restricted. In such a culture, sharing data stories and establishing controlled data-sharing forums are beneficial for promoting open communication (Wolff et al., 2016).

To design effective data training programs, organizations must first identify the specific data literacy needs and competencies required for each data culture type. The unified framework proposed by Wolff et al. (2016) encompasses competencies such as understanding, collecting, analyzing, summarizing, interpreting, and prioritizing data. In a Progressive data culture, leaders should focus on providing advanced data analytics training to employees. For a Traditional data culture, data interpretation training is crucial for developing confidence in data-driven decision-making. In a Protectionist data culture, the emphasis should be on understanding data security measures and risk management to build trust in data-driven decision-making. In contrast, a Preservationist data culture requires training employees to leverage data insights effectively for continuous improvement (Gummer & Mandinach, 2015).

To successfully implement data training programs, organizations must foster a data-driven culture that encourages continuous learning and skill development. MIT Sloan (2023) emphasizes that leaders must take responsibility for promoting data literacy and recognizing the broader data strategy. Leaders should actively engage in data projects to gain practical experience and understanding, emphasizing how data initiatives lead to financial returns and improved decision-making. For example, in a Preservationist data culture, displaying successful data-driven projects and initiatives can drive data-driven behavior and reinforce the organization's core values.

Implementing data training programs may face challenges and resistance. The State of Data Literacy in 2023 report (Nehme, 2023) highlights challenges such as lack of budget, inadequate training resources, and employee resistance. To address these challenges, organizations must gain executive support and align learning objectives with business goals. In a Protectionist data culture, addressing concerns and fears associated with data-driven initiatives is crucial. Effective communication and setting clear expectations about the use of data can alleviate resistance.

Measuring the effectiveness of data training is essential to assess the impact and success of the programs. Nehme (2023) suggests tracking a variety of ROI signals and regularly assessing skill acquisition over time. The report emphasizes the need for continuous upskilling, culture change, and transforming organizational habits to bridge the data skills gap effectively. Designing and implementing data training programs and resources is a critical aspect of fostering a data-literate workforce and promoting a data-driven culture within organizations. By understanding the unique characteristics of each data culture type, identifying user needs and competencies, and overcoming challenges and resistance, organizations can create effective data training initiatives. Continuous measurement and assessment of the training's effectiveness are vital to ensure long-term success and enable organizations to stay competitive in an increasingly data-driven world.

Continuous Learning and Data Skills

Encouraging a continuous learning culture is paramount for organizations to enhance data skills and capabilities. A data-literate workforce empowers individuals at all levels of an organization to work effectively with data. In this section, we explore the importance of fostering a continuous learning culture and how it can be tailored to different data culture types based on the Usage and Flow Data Culture Model.

To build and maintain data literacy within an organization, leadership must recognize that data skills are not static but need to be continuously cultivated. As highlighted by Azhar (2022), data literacy is a spectrum of fluency that encompasses a range of data skills. Organizations must understand that data literacy is not limited to the analytics or IT team but is critical for all departments and roles. A continuous learning culture fosters a mindset of curiosity, adaptability, and growth, allowing employees to stay updated with the latest data practices and technologies. According to the article by Ghosh (2021), the skills economy, particularly the data skills economy, plays a decisive role in an organization's success and overall performance.

In a Progressive data culture, Data Usage is advanced, and data flow is unrestricted. To encourage continuous learning, organizations should provide advanced data analytics training and opportunities for employees to delve into complex data-driven insights (Gummer & Mandinach, 2015). Leaders must actively engage in data projects to gain practical experience and understanding (MIT Sloan, 2023). A data-driven organization with a Progressive data culture should foster a culture of experimentation and risk-taking. Continuous learning should be encouraged through regular workshops, seminars, and collaborative projects that allow employees to innovate with data and explore new possibilities (Ghosh, 2021).

In a Traditional data culture, Data Usage is limited, but data flow is democratized. Encouraging continuous learning in this context requires providing data interpretation training to employees (Gummer & Mandinach, 2015). Organizations should prioritize developing personalized learning paths that align with different data personas (Nehme, 2023). A data-driven organization with a Traditional data culture can promote cross-functional learning and knowledge exchange. By breaking down silos and providing opportunities for employees from different departments to collaborate on data projects, the organization fosters a culture of continuous learning and data-driven decision-making.

In a Protectionist data culture, Data Usage is basic and data flow is limited. To encourage continuous learning, organizations should promote transparent communication around Data Usage. A data-driven organization with a Protectionist data culture should prioritize building trust in data-driven decision-making through continuous education on data security measures and risk management (Wolff et al., 2016). Regular workshops on data privacy, data-handling techniques, and data ethics can help employees feel more comfortable with Data Usage.

In a Preservationist data culture, Data Usage is advanced, but data flow is restricted. To foster continuous learning, organizations should encourage employees to share data-driven stories that provide valuable insights. Controlled data-sharing forums can be established to facilitate secure data collaboration (Wolff et al., 2016). A data-driven organization with a Preservationist data culture should emphasize the importance of continuous improvement through data insights. Continuous learning opportunities can focus on how data-driven practices align with the organization's heritage and long-term objectives (Gummer & Mandinach, 2015).

To sustain a continuous learning culture, organizations must provide resources and support to employees. As emphasized by the article by Nehme (2023), change management and skills development are necessary when bridging the data skills gap. Organizations can offer a mix of instructor-led sessions, online training, and gamified learning experiences to cater to different learning preferences. Providing access to curated learning paths and interactive resources can help employees navigate the vast array of data-related content and apply their skills effectively.

Encouraging a continuous learning culture enhances data skills and capabilities within an organization. By recognizing the significance of data literacy as a spectrum of fluency, tailoring ongoing learning initiatives to different data culture types, and providing resources and support, organizations can create a data-driven workforce that is adaptable, innovative, and ready to thrive in the dynamic data landscape. A continuous learning culture empowers employees to make data-driven decisions and drive success for the organization in an increasingly data-centric world.

Data Literacy in a Traditional Culture

In a Traditional culture, which represents organizations with democratized Data Flow and basic Data Usage, according to the UFDCM, data literacy and skill development may have a different focus compared to a Progressive culture. Let's explore what data literacy and skill development might look like in a Traditional culture.

1. Basic Data Literacy Training

   In a Traditional culture, data literacy training may primarily focus on providing employees with basic knowledge and understanding of data concepts. This includes familiarizing employees with common data terminology, basic data analysis techniques, and the ability to interpret simple data visualizations. The emphasis is on equipping employees with foundational data literacy skills to understand and navigate data-related discussions and reports.

2. Practical Data Skills

    Skill development in a Traditional culture may lean toward
    practical data skills that are directly relevant to employees' day-to-
    day tasks. This may involve training employees on specific tools
    and software commonly used in their roles, such as spreadsheet
    applications or industry-specific data management systems. The
    goal is to enhance employees' ability to work with data and react
    to trends rather than using advanced analytics.

3. Limited Data Communication

    In the Traditional culture, data communication is focused on
    measurement and diagnostics with limited predictive analytics.
    Very little proactive discovery and communication happen. The
    objective is to provide information for decision-making rather
    than emphasizing visual storytelling or data-driven narratives.

4. Limited Continuous Learning

    In a Traditional culture, the emphasis on continuous learning
    and development may be lower as compared to a Progressive
    culture. Training initiatives may be more ad hoc and focused on
    addressing immediate needs rather than promoting a consistent
    culture of learning. Resources for ongoing skill development, such
    as external courses or mentorship programs, may be limited or
    less accessible.

Data Literacy in a Protectionist Culture

In a Protectionist culture, which represents organizations with limited Data Flow and
basic Data Usage according to the Usage and Flow Data Culture Model (UFDCM), data
literacy and skill development may face certain challenges and limitations. However,
there may still be opportunities for data literacy and skill development within this
context. Let's explore what data literacy and skill development might look like in a
Protectionist culture.

1. Limited Data Accessibility

    In a Protectionist culture, the emphasis is on tightly controlling
    access to data and limiting information flow. This restricted Data
    Flow may result in limited opportunities for employees to directly
    engage with and develop data literacy skills. Access to data may
    be limited to a select few individuals or departments, and there
    may be a lack of transparency and collaboration in data-related
    processes.

2. Specialized Training Programs

    Due to the controlled data environment, organizations in a
    Protectionist culture may implement specialized training
    programs to develop data literacy skills among a specific group of
    individuals responsible for data management and analysis. These
    programs may focus on technical skills such as data collection,
    data cleansing, and basic data analysis techniques. Because of its
    nature, access control concepts such as security clearances may
    be required for specific data domains.

3. Centralized Decision-Making

    In a Protectionist culture, decision-making tends to be
    centralized, with a limited number of individuals or departments
    responsible for making data-driven decisions. This centralized
    approach may limit opportunities for employees to develop data
    literacy skills and contribute to data-driven decision-making
    processes.

4. Focus on Data Governance and Privacy

    Protectionist cultures often prioritize data governance and privacy
    measures to protect sensitive data. Therefore, skill development
    efforts may focus on ensuring compliance with data protection
    regulations, understanding data governance frameworks, and
    implementing privacy policies.

Data Literacy in a Preservationist Culture

In a Preservationist culture, which represents organizations with limited Data Flow but advanced Data Usage according to the Usage and Flow Data Culture Model (UFDCM), data literacy and skill development may take on a unique character. Let's explore what data literacy and skill development might look like in a Preservationist culture.

1. Emphasis on Data Stewardship

    In a Preservationist culture, where data is highly valued and protected, data literacy and skill development may focus on cultivating a sense of responsibility and stewardship toward data assets. Employees may receive training on data governance, data quality management, and data security to ensure the preservation and integrity of data.

2. Controlled Data Access

    Preservationist cultures prioritize data protection, which may result in restricted access to data. Employees may undergo training programs to understand the protocols and procedures for accessing and utilizing data within the established boundaries and guidelines. This training can include data access controls, data-handling procedures, and compliance with privacy regulations.

3. Compliance and Risk Management

    Preservationist cultures often place a strong emphasis on compliance and risk management when it comes to Data Usage. Data literacy and skill development efforts may involve training employees on legal and ethical aspects of Data Usage, as well as understanding the potential risks associated with data-driven decision-making.

4. Specialized Data Analysis Skills

    In a Preservationist culture, where Data Usage is advanced, employees may receive training in specialized data analysis skills. This may include advanced statistical techniques, predictive modeling, and data visualization to extract insights from the available data while adhering to the established data access and usage protocols. Although sharing data flow is limited, the sharing of data analytics techniques will play a role in data literacy and skill development.

Data Literacy in a Progressive Culture

In a Progressive culture, which represents organizations with democratized Data Flow and advanced Data Usage according to the Usage and Flow Data Culture Model (UFDCM), data literacy and skill development are fundamental pillars for driving innovation, collaboration, and data-driven decision-making. Let's explore what data literacy and skill development might look like in such a culture.

1. Emphasis on Data Literacy at All Levels

    In a Progressive culture, data literacy is promoted among employees at all levels of the organization. There is a recognition that data literacy should not be limited to data specialists but should be a pervasive skill set throughout the organization. This includes equipping employees with the ability to understand data, interpret insights, and effectively communicate with data.

2. Comprehensive Data Training Programs

    To foster data literacy, organizations in a Progressive culture prioritize the design and implementation of comprehensive data training programs. These programs aim to enhance employees' data skills and capabilities, enabling them to work with data effectively. Training initiatives may include a variety of approaches, such as interactive workshops, online courses, hands-on projects, and mentorship opportunities.

3. Data Visualization and Communication Skills

    In a Progressive culture, there is an emphasis on developing data visualization and communication skills. Employees are encouraged to present data insights in a visually compelling and understandable manner to facilitate decision-making and drive action. This involves training employees on data visualization tools, storytelling techniques, and effective communication strategies to ensure data is conveyed accurately and compellingly.

4. Continuous Learning and Development

    A Progressive culture promotes a continuous learning mindset, where employees are encouraged to continuously develop and refine their data skills. This involves providing ongoing learning opportunities, such as access to external resources, communities of practice, and mentorship programs. Organizations also foster a culture that values learning from data-driven successes and failures, encouraging employees to reflect on their experiences and continuously improve their data literacy and skills.

Triggers for Data Literacy

Regardless of the data culture selected, the organization should put into place triggers for training events. These triggers will account for the life cycle of the employee, the life cycle of the organization, and the life cycle of data. As employees, data, and organizations progress through these life cycles, associated data literacy activities should follow.

Data literacy activities are necessary throughout the life cycle of the employee. When an employee is hired, the employee should receive organizational and job-specific data literacy training. Organizational training includes the most current regulatory, compliance, and security training. Additionally, policies for data handling, business nomenclature, industry terminology, etc. should be provided to new employees. The new hire will also receive data literacy for their specific job area.

Another phase of the employee life cycle is the promotion or transfer phase. With an employee promotion or transfer, the potential exists for new or different data or tools to be used by the employee. Data literacy training should trigger training requirements for the employee to ensure this knowledge is received. The data literacy program must account for the employee life cycle.

The organizational life cycle includes the organization's maturity, industry considerations, and regulatory and compliance efforts. As the organization matures from a startup to a growing entity, the organization may require some refinement of data accessibility, new data sources, or increased use of other tools. Industries continually change as well, and when data points are created or modified, the data literacy program must integrate this learning and educate the employee. The same is true of regulatory or compliance requirements. As regulatory bodies update requirements, the data literacy program must respond promptly.

The life cycle of data is important as well. As data ages or data retention reaches its maturation, the data may be anonymized, obfuscated, or masked. As data moves through the life cycle or time horizons change throughout the ecosystem, the data literacy program must trigger additional training to increase data efficacy and decision-making.

Summary

In this chapter, we cover the important concept of data literacy within the context of fostering a data culture in organizations. In today's data-driven world, the value of data is widely acknowledged, making data literacy a vital skill for individuals at all levels of an organization. This chapter explores the significance of data literacy and its implications for organizations striving to harness the power of data effectively.

Data literacy is the ability to comprehend, assess, and work with data proficiently. It encompasses skills beyond technical expertise, including the capacity to interpret data, ask relevant questions, analyze information, and derive actionable insights. Data literacy involves understanding the data's context, limitations, and ethical implications. Data-literate individuals make decisions based on evidence and insights, leading to improved outcomes and mitigating risks. Data literacy fosters collaboration between data experts and domain specialists, promoting effective communication and sharing of insights.

Building a data culture requires intentional efforts to promote data literacy across an organization. Organizations should provide data literacy training for employees at all technical levels, focusing on critical thinking, interpretation, and communication. Every effort should be made to identify and empower data literacy advocates to bridge gaps between technical and nontechnical staff. Additional emphasis should be placed on data visualization and storytelling techniques for effective data communication. Accessible data tools and technologies help promote a culture of learning by providing ongoing opportunities to enhance data skills.

Organizations should design training programs tailored to their data culture type, whether Traditional, Protectionist, Preservationist, or Progressive, as outlined in the UFDCM. Training should align with the competencies required for each type, encompassing understanding, analyzing, interpreting, and prioritizing data. A continuous learning culture is crucial for enhancing data skills. Different data culture types require specific approaches to continuous learning, such as advanced analytics for Progressive cultures, data interpretation for Traditional cultures, transparency for Protectionist cultures, and sharing data stories for Preservationist cultures.

Data literacy triggers should be aligned with employee, organization, and data life cycles. Training should occur during onboarding, promotions, regulatory changes, and changes in data flow or usage. Data literacy forms the bedrock of a successful data culture. It empowers individuals to engage with data confidently. By promoting data literacy through comprehensive training, continuous learning, and tailored approaches for different data cultures, organizations can fully harness the potential of data and establish a culture of data-driven excellence.

CHAPTER 7

Embedding Data into Decision-Making

Edward R. Tufte, in *The Visual Display of Quantitative Information*, is quoted as saying, "Above all else show the data." In today's data-driven landscape, the successful cultivation of a robust data culture has become a paramount goal for organizations seeking to gain a competitive edge and navigate the complexities of the modern business environment. This chapter discusses two intertwined aspects that play a pivotal role in shaping a thriving data culture: data analytics and data privacy/security. By examining these aspects through the lens of the Usage and Flow Data Culture Model and considering the distinct culture types it encompasses—Protectionist, Preservationist, Traditional, and Progressive—we gain a comprehensive understanding of how data analytics and data privacy/security contribute to building a data culture that is not only efficient and informed but also ethically responsible.

The foundation of a data culture rests on the ability to derive meaningful insights from the vast reservoirs of data available within an organization. As data analytics permeates every facet of decision-making, it aligns closely with the understanding-oriented Data Usage highlighted in the UFDCM. The Protectionist quadrant, characterized by its cautious approach toward Data Usage, seeks to ensure that data analytics adhere to stringent protocols, safeguarding against potential risks. Preservationist organizations, concerned with retaining the integrity of data and continuous improvement, utilize analytics to glean historical insights that contribute to their long-term vision.

In the Traditional quadrant, data analytics supports established processes and aids in optimizing familiar workflows. Here, data-driven insights are often leveraged to streamline operations and enhance existing practices. In contrast, the Progressive quadrant, driven by a thirst for innovation, harnesses the power of cutting-edge analytics techniques to uncover novel patterns and opportunities. By understanding the

© Gary W. Griffin, David Holcomb 2023
G. W. Griffin and D. Holcomb, *Building a Data Culture*, https://doi.org/10.1007/978-1-4842-9966-1_7

alignment between data analytics and these different culture types, organizations can strategically mold their approach to analytics adoption, catering to their unique needs and priorities.

Data analytics transcends raw numbers to deliver actionable insights, enabling organizations to make informed decisions. These insights resonate profoundly with the understanding-oriented Data Usage prevalent in a Preservationist culture. Through the careful analysis of historical data, Preservationist organizations gain insights that bolster their commitment to tradition and legacy. Protectionist entities, equally focused on the depth of insights, emphasize responsible and secure analytics practices to ensure data remains protected.

In the Traditional quadrant, data analytics refines existing processes, optimizing efficiency and facilitating data-backed decision-making. Here, analytics offers a pragmatic approach, enabling incremental improvements in established practices. In the Progressive quadrant, integrating advanced analytics techniques fuels innovation and disruptive thinking. By embracing predictive modeling, machine learning, and other sophisticated tools, Progressive organizations push boundaries and expand the horizons of possibility.

As data analytics thrives, the flow of data becomes essential, aligning with the Data Flow vector in the UFDCM. In the Protectionist quadrant, where controlled data access is highly valued, robust data governance ensures that analytics efforts remain compliant and secure. Preservationist organizations, while valuing data preservation, find a balance between data access and data protection, enabling analytics to serve as a bridge between traditional and modern insights. In the Traditional quadrant, data flow optimization streamlines processes, ensuring that relevant data is readily available for analysis. Here, data pipelines are designed to enhance the efficacy of established practices, demonstrating the symbiotic relationship between data flow and analytics. In the Progressive quadrant, open data flows facilitate the exploration of uncharted territories, enabling data scientists and analysts to push the boundaries of innovation.

In this chapter, we embark on a journey through the intricate tapestry of data analytics and data privacy/security, guided by the Usage and Flow Data Culture Model. By understanding how these two pillars interact with the distinct cultural attributes of Protectionist, Preservationist, Traditional, and Progressive organizations, we uncover the blueprint for building a data culture that harnesses the power of analytics while safeguarding the integrity of data privacy and security. Through this exploration, we lay the groundwork for an enlightened and ethically sound data culture that thrives in an era where data insights reign supreme.

Data Analytics and the Data Culture Model

Within the UFDCM, data analytics assumes a pivotal role in fostering a thriving data-driven culture across organizations. Specifically, advanced analytics acts as the pinnacle of the Data Usage vector to realize understanding and differentiation. The UFDCM relies heavily on concepts of data leadership, data governance, and data adoption to guide the integration of data analytics practices.

At the heart of the UFDCM lies robust data leadership, which emphasizes the significance of data-driven decision-making (Harbert, 2021). Data analytics significantly contributes to this dimension by enabling leaders to extract meaningful insights from vast datasets, thereby guiding strategic planning, setting informed goals, and monitoring organizational progress. Effective data governance plays a critical role by ensuring the quality, consistency, and security of organizational data (Twilio Inc., 2023). Data analytics is instrumental in upholding data governance practices by offering insights into data quality, usage patterns, and potential risks. It contributes to maintaining high data standards through processes like data cleaning, transformation, and validation.

Data adoption focuses on creating a culture where employees embrace data-driven decision-making at all levels (Rushin, 2022). Data analytics strengthens this dimension by democratizing data access and analysis. Self-service analytics tools empower employees to explore data, gain insights, and make informed decisions without requiring specialized technical expertise (Murguia, 2022). This approach enhances data literacy, encourages collaboration, and drives innovation.

By integrating data analytics across these dimensions, organizations cultivate a data culture that propels innovation, enables informed decision-making, and maximizes business value (Abdul-Jabbar et al., 2022; Stodder, 2017; Calzon, 2023). As organizations endeavor to establish and nurture a data-driven culture, data analytics acts as the bridge that connects the visionary potential of data with tangible, impactful outcomes.

Leveraging Data Analytics for Insights

Within the context of the UFDCM, the "Data Usage" vector holds particular significance. This vector revolves around the utilization of data to extract insights and facilitate informed decision-making, with a specific focus on understanding-oriented Data Usage. The alignment between data analytics and the "Data Usage" vector of the UFDCM highlights the importance of data analytics for gaining insights and making informed decisions. Specifically, the understanding-oriented Data Usage aspect of this vector emphasizes the value of meaningful interpretation and comprehension of data, rather than data accumulation for its own sake. This aligns perfectly with the broader goals of the Usage and Flow Data Culture Model, which aims to cultivate an organizational environment where data-driven insights drive strategy and operations.

Data analytics serves as a critical enabler of understanding-oriented Data Usage. It involves employing advanced analytical techniques to uncover hidden patterns, correlations, and trends within vast datasets. Such insights offer a deeper understanding of various aspects of the business, including customer behavior, market dynamics, and operational efficiency (Abdul-Jabbar & Farhan, 2022). For instance, the transformative power of data analytics is exemplified in Elizabeth Puchek's experience as the former CDO at the U.S. Citizenship and Immigration Services, where data analysis led to significant process improvements and organizational credibility (Harbert, 2021).

Informed decision-making is a cornerstone of effective leadership, and data analytics plays an essential role in this process. By translating raw data into actionable insights, data analytics empowers decision-makers with evidence-based information, reducing reliance on intuition and guesswork. This is highlighted in the case of Private Equity professionals, who can significantly enhance their investment strategies by harnessing data analytics to drive strategic decisions (Performance Improvement Partners, n.d.).

In the dynamic landscape of modern business, the role of data analytics has emerged as a driver of insights, strategic decision-making, and organizational growth within the context of the UFDCM. The multifaceted realm of data analytics encompasses techniques such as data visualization, data mining, and machine learning. It further demonstrates how organizations harness data analytics to create value and elevate their decision-making processes, highlighting the relationship between a mature data culture and the strategic adoption of data analytics.

Tufte also said, "Graphical excellence is that which gives to the viewer the greatest number of ideas in the shortest time with the least ink in the smallest space." One of the fundamental data analytics techniques is data visualization, which transforms complex datasets into comprehensible visual representations. These visuals, ranging from charts and graphs to interactive dashboards, facilitate a clear understanding of trends, patterns, and outliers (Murguia, 2022). Visualizations allow decision-makers to grasp insights swiftly, leading to informed actions and improved business outcomes.

Data mining involves the systematic exploration of large datasets to uncover hidden patterns and relationships. It enables organizations to identify correlations, predict future trends, and extract valuable knowledge from existing data (Abdul-Jabbar & Farhan, 2022). By applying data mining techniques, companies can optimize marketing campaigns, enhance customer experiences, and streamline operational processes. Machine learning, a subset of artificial intelligence, empowers systems to learn and improve from experience without explicit programming. Organizations leverage machine learning algorithms to develop predictive models, automate tasks, and gain deeper insights from data (Harbert, 2021). For instance, machine learning algorithms can analyze historical sales data to forecast future demand, aiding inventory management.

Here are a few examples of data analytics in action:

> Driving Customer Engagement: Emami, a personal care and healthcare business, harnessed data analytics to track financial and operational metrics using tailored visualizations (Murguia, 2022). This approach empowered Emami to engage with customers more effectively, customize offerings, and bolster customer satisfaction.

> Enhancing Efficiency: The U.S. Citizenship and Immigration Services significantly reduced processing time by improving a data-transfer process through data analytics (Harbert, 2021). By identifying bottlenecks and causes of backlogs, the organization streamlined operations and improved its mission fulfillment.

> Optimizing Investments: Private Equity professionals utilize data analytics to gain insights into investment opportunities, enhancing their investment strategies and maximizing returns (Performance Improvement Partners, n.d.). Data-driven decision-making guides their portfolio management, ensuring prudent investments.

In a mature data culture, the strategic adoption of data analytics is emblematic of an informed and empowered workforce. Such a culture encourages employees to embrace data analytics as a tool for problem-solving, innovation, and evidence-based decision-making. The alignment between data analytics and a mature data culture propels organizations toward agility and competitiveness, where insights from data analytics become a driving force for growth.

Data analytics is directly linked to the "Data Usage" vector of the UFDCM. It serves as a catalyst for understanding-oriented Data Usage, enabling organizations to unlock valuable insights, make informed decisions, and shape their strategies based on empirical evidence. By embracing techniques such as data visualization, data mining, and machine learning, organizations can harness the power of data to derive insights, drive value, and enhance decision-making processes. Data analytics contributes to customer engagement, efficiency enhancements, and optimized investments, all while nurturing a mature data culture that fosters informed and agile decision-making.

Balancing Data Flow for Effective Analytics

Achieving an equilibrium between data accessibility and managing information overload is a critical aspect of fostering a data-driven culture. The intricate interplay between providing access to relevant data and avoiding the pitfalls of overwhelming data abundance is examined. It highlights the vital role of data governance in regulating data flow, ensuring the availability of necessary information, and maintaining data quality. Additionally, the section presents strategies that organizations can adopt to optimize data flow, thereby supporting analytical endeavors while safeguarding against potential distractions.

In the modern business landscape, the availability of copious amounts of data has transformed decision-making processes. However, the challenge lies in striking the right balance between granting access to relevant data and preventing information overload (Twilio Inc., 2023). While data-driven insights are invaluable, an excess of unfiltered data can hinder efficient decision-making, leading to confusion and cognitive overload.

Effective data governance emerges as a cornerstone in managing the delicate balance of data flow. Data governance involves establishing and enforcing policies, processes, and guidelines for data management (Stodder, 2017). By implementing robust data governance practices, organizations ensure that data is accurate, consistent, and available to those who need it. This regulatory framework helps prevent data chaos, ensuring the right data is accessible for analytical purposes.

The following strategies work well to support analytical purposes based on Data Flow:

1. Tailored Data Accessibility: Organizations can optimize data flow by tailoring data accessibility to specific user roles and responsibilities (Twilio Inc., 2023). By providing individuals with access to data directly relevant to their roles, organizations can prevent the dilution of focus and mitigate the risk of information overload.

2. Automated Data Quality Checks: Implementing automated data quality checks as part of data governance practices ensures that the data flowing into analytical processes is accurate and reliable (Stodder, 2017). This reduces the likelihood of erroneous insights derived from flawed data. Coupling these automated quality checks with proactive data quality initiatives such as master data management increases user involvement, data quality, and trust.

3. Data Cataloging and Metadata Management: Data cataloging and metadata management enable users to quickly discover and understand available data assets, enhancing the efficiency of data retrieval and analysis (Rushin, 2022). By facilitating data discovery, organizations can promote informed decision-making without overwhelming users with unnecessary data.

Within a data-driven culture, individuals are empowered to become discerning consumers of data. They are equipped with the skills and knowledge to identify and access the data necessary for their tasks, fostering a culture of efficient data utilization (Calzon, 2023). This active engagement with data, coupled with the guidance of data governance principles, ensures that data flow remains optimized, promoting effective analytics and informed decision-making.

The data culture model championed in this book underscores the delicate nature of balancing data flow for effective analytics. By appreciating the significance of data governance, organizations can regulate data access, prevent information overload, and optimize data flow. Through strategies such as tailored data accessibility, automated data quality checks, and data cataloging, organizations can navigate the complexities of data abundance while nurturing a culture where data is harnessed purposefully to drive insights and value.

Data Privacy and Security Within the Data Culture Model

When activating the data asset to enhance decision-making, the organization must embed robust data privacy and security measures into its data culture. The "Data Flow" vector of the UFDCM, which involves the smooth movement of data across the organization, is the focal point for implementing these measures. This section explores the alignment of data privacy and security within the "Data Flow" vector of the UFDCM and emphasizes their pivotal role in protecting sensitive information and maintaining trust.

The "Data Flow" vector within the UFDCM encompasses the entire journey of data, from its generation and collection to its transformation, analysis, and dissemination. At each stage of this journey, data privacy and security considerations are vital to ensure the confidentiality, integrity, and availability of the information (Martin, 2019). An examination of various articles sheds light on the multifaceted aspects of data privacy and security within the context of this vector.

Komnenic's (2022) exploration of data privacy compliance emphasizes how adherence to data privacy principles and regulations contributes to the establishment of trust between businesses and consumers. The alignment of data privacy and security practices within the "Data Flow" vector is evident as organizations must safeguard data at every stage to build and maintain this trust. As data traverses the organization, robust data protection mechanisms must be in place to prevent breaches and unauthorized access (Shan et al., 2021).

Data privacy and security are essential in protecting sensitive information, such as personally identifiable information (PII) and protected health information (PHI). The article by Cloudian (2023) emphasizes the significance of data protection in safeguarding such sensitive data from breaches, reputation damage, and regulatory noncompliance. Within the "Data Flow" vector, the implementation of access control models, encryption techniques, and other security measures ensures that sensitive data remains confidential and is accessed only by authorized personnel (Martin, 2019).

Furthermore, establishing a culture of privacy, as discussed by Weller and Leach (2020), is integral to maintaining trust with customers and employees. Organizations must instill a mindset where data privacy and security are deeply ingrained in the organizational culture. This cultural shift aligns with the "Data Flow" vector by promoting responsible data-handling practices at every stage.

Data privacy and security are inseparable components of the "Data Flow" vector within the UFDCM when building a data culture. As data journeys through the organization, it is imperative to ensure robust data protection measures are in place to safeguard sensitive information. Adhering to data privacy principles, implementing access controls, and fostering a culture of privacy are essential steps toward maintaining trust, preventing breaches, and upholding the integrity of the data flow.

Establishing Robust Data Privacy Measures

Within the UFDCM, the "Data Flow" vector emerges as a fundamental pillar associated with data privacy, focusing on safeguarding sensitive data through well-defined policies, access controls, and advanced security techniques. The decisions made in the "Data Flow" vector will affect the "Data Usage" vector by regulating the data available and its proper use rules. This section delves into the significance of data privacy policies and procedures, the importance of restricting data access, and the implementation of key data privacy measures.

At the heart of a robust data culture lies a meticulous framework of policies and procedures designed to safeguard sensitive information. Qureshi (2020) underscores the need for a privacy culture that aligns with an organization's values and integrates privacy principles into everyday practices. Such policies not only ensure compliance with data protection regulations but also set the tone for responsible data handling across the organization. These policies outline the permissible uses of data, dictate retention periods, and establish protocols for data sharing, minimizing the risk of unauthorized disclosures and data misuse.

The core of data privacy lies in restricting access to authorized individuals while taking measures to prevent data breaches. The UFDCM's "Protectionist" data culture type aligns seamlessly with this notion, emphasizing the importance of limiting data exposure only to those requiring it. Termly (2023) highlights the significance of maintaining confidentiality and preventing breaches, underscoring the financial, legal, and reputational implications when data is compromised.

A critical aspect of establishing robust data privacy measures is the implementation of effective techniques and tools. Encryption stands out as a potent shield against unauthorized access, rendering data indecipherable to those without the appropriate decryption keys (Cloudian, 2023). Access controls, a fundamental element of the "Data Flow" vector, delineate who can access specific data and under what conditions (Martin,

2019). Additionally, user authentication serves as a gateway to data, ensuring that only authorized personnel gain entry (Termly, 2023). These measures collectively contribute to a layered defense, fortifying the organization's data privacy stance.

In the journey to foster a data culture that thrives on integrity and security, the "Flow" vector occupies a principal role. With thorough policies and procedures, the organization can cultivate responsible data-handling practices. The restricted access paradigm ensures that data remains in the hands of authorized personnel, bolstering confidentiality. Meanwhile, the strategic implementation of data privacy measures such as encryption, access controls, and user authentication adds multiple layers of protection. By embracing this vector, organizations forge a path toward a data culture that optimizes data utilization and champions data privacy and security.

Impact of Data Privacy and Security on Data Usage

Although much of data privacy and security measures are implemented in the Flow vector, the "Data Usage" vector will be affected. The "Data Usage" vector forms the foundation of responsible and effective data exploration and analysis. This section delves into the intricate relationship between data privacy, security, and employee confidence in Data Usage. It highlights the role of stringent security measures in facilitating active engagement with data and underscores the ethical and legal imperatives of ensuring compliance with data privacy regulations.

The nexus between data privacy, security, and employee confidence in data exploration and analysis is unequivocal. Employees are more likely to engage confidently with data when they are assured of its privacy and security. Data privacy aptly encapsulates the nature of this relationship by emphasizing the value of transparency and responsible data practices (Qureshi, 2020). Studies have shown that individuals are more willing to participate actively in data-driven initiatives when they trust that their sensitive information is handled ethically and securely (Termly, 2023).

The integration of appropriate security measures into an organization's data culture empowers employees to actively engage with data. By ensuring that robust data security controls are in place, organizations create an environment that encourages data exploration and analysis without fear of data breaches or unauthorized access. The "Protectionist" data culture type resonates with this concept, highlighting the significance of controlled and authorized data sharing (Martin, 2019). This controlled access not only safeguards sensitive information but also fosters a sense of ownership and empowerment among employees, enabling them to harness data-driven insights confidently.

Data privacy, security, and ethical Data Usage have an inseparable connection. Organizations must adhere to data privacy regulations to ensure that Data Usage aligns with ethical and legal standards. As described in Chapter 5, data governance within each data culture type exemplifies the systematic approach needed for regulatory compliance (Dennis, 2020). Compliance with regulations such as GDPR, CCPA, and HIPAA reinforces the ethical foundation of Data Usage by respecting individuals' privacy rights and preventing misuse (Termly, 2023).

The "Data Usage" vector encapsulates the intricate interplay between data privacy, security, employee confidence, and ethical Data Usage. By fostering a culture of data privacy and security, organizations not only empower employees to explore and analyze data actively but also ensure compliance with ethical and legal standards. Employee confidence flourishes in an environment where stringent security measures protect sensitive information, enabling data-driven decision-making without compromise. The integration of these principles within the data culture sets the stage for responsible and impactful Data Usage, driving organizations toward their strategic objectives.

Building Trust and Democratization Through Security

The Usage and Flow Data Culture Model (UFDCM) explains the counterintuitive way security can be used to enhance data flow and democratization rather than impede it. This section examines the critical interplay between data privacy, security, and the broader goals of building trust and democratization through security measures. By exploring the relationship between strong data privacy and security, trust building, open data sharing, and data democratization, organizations can create a data-driven ecosystem that fosters collaboration, innovation, and transparency.

Trust is the cornerstone of any successful data culture. The UFDCM emphasizes the significance of trust building through transparent and ethical data practices (Qureshi, 2020). Robust data privacy and security measures are pivotal in establishing and nurturing this trust. Stakeholders, including customers, partners, and employees, are more likely to engage confidently with data when they are assured that their sensitive information is safeguarded and used responsibly (Termly, 2023). Organizations prioritizing data security protect against breaches and send a powerful message about their commitment to respect privacy rights and uphold ethical standards.

Contrary to the misconception that data security hinders data democratization, strong security measures can enable and promote a culture of open data sharing. The "Progressive" data culture type in the UFDCM underscores the value of data accessibility and availability for informed decision-making (Termly, 2023). When stakeholders have confidence in the security of data-sharing mechanisms, they are more likely to contribute, collaborate, and leverage data to drive strategic outcomes. Secure data-sharing platforms and technologies empower individuals to access, analyze, and derive insights from data without compromising privacy or security.

The constructive interaction between security and transparency is necessary for striking the right balance between data accessibility and protection. The "Protectionist" data culture type recognizes the importance of controlled and authorized data sharing (Martin, 2019). Security measures, such as encryption, user authentication, and access controls, ensure data is accessible only to authorized individuals. Simultaneously, transparent data practices, combined with strong security protocols, foster an environment of accountability and responsibility. By embracing this balance, organizations enhance data accessibility while upholding data privacy and security standards.

The "Data Usage" vector relies on the integral relationship between data privacy, security, trust building, open data sharing, and democratization. By prioritizing strong data privacy and security measures, organizations instill trust among stakeholders and create an environment where data is harnessed collaboratively, innovatively, and responsibly. The strategic integration of security mechanisms bolsters data accessibility, empowers stakeholders, and paves the way for a thriving data culture where the potential of data-driven insights can be fully realized.

Analytics, Privacy, and Security in a Traditional Data Culture

Within a Traditional data culture, the orchestration of data analytics, data privacy, and security assumes a distinctive character, reflecting a conservative and structured approach to data utilization. In the UFDCM, a Traditional data culture places a premium on data stewardship, controlled dissemination, and prudent decision-making.

Data Analytics in a Traditional Data Culture

In a Traditional data culture, data analytics are harnessed with a cautious and methodical mindset. Data is regarded as a valuable resource to be curated and analyzed by designated experts. Analytical processes adhere to established protocols, emphasizing accuracy and reliability. Analytics initiatives are driven by predetermined objectives, aligning closely with the organization's strategic goals.

Strategies for Data Analytics in a Traditional Data Culture

1. Centralized Analytics: Centralize data analytics efforts under a dedicated team of data experts who conduct thorough analyses and provide insights to key stakeholders.

2. Structured Reporting and Dashboards: Emphasize structured reporting, dashboarding, and visualization formats that adhere to standardized templates and methodologies, ensuring consistency and accuracy in data presentation.

3. Data Governance Framework: Develop a robust data governance framework that outlines data quality standards, data ownership, and Data Usage policies to guide analytics endeavors.

Data Privacy in a Traditional Data Culture

In a Traditional data culture, data privacy is approached with a cautious and compliance-oriented perspective. Individual rights and regulatory frameworks are key considerations. Data sharing is tightly controlled, and privacy measures are precisely applied to prevent unauthorized access to sensitive information.

Strategies for Data Privacy in a Traditional Data Culture

1. Explicit Consent: Prioritize obtaining explicit consent from individuals before using their data, adhering to legal requirements and ethical considerations.

2. Data Minimization: Collect only the minimum amount of data necessary for specific purposes, reducing the risk associated with storing excessive personal information.

3. Regular Audits: Conduct regular privacy audits to ensure compliance with privacy regulations, assess data-handling practices, and identify areas for improvement.

Data Security in a Traditional Data Culture

Security in a Traditional data culture is characterized by stringent controls and well-defined access privileges. Data is carefully compartmentalized, and security protocols are implemented to safeguard data integrity and prevent unauthorized breaches.

Strategies for Data Security in a Traditional Data Culture

1. Role-Based Access Control (RBAC): Implement RBAC to ensure data access is granted based on defined roles and responsibilities, minimizing the risk of data exposure.

2. Encryption: Employ encryption techniques to protect data both at rest and during transmission, ensuring that sensitive information remains confidential.

3. Access Logs and Monitoring: Maintain detailed access logs and implement continuous monitoring to track data access activities and detect potential security breaches.

In a Traditional data culture, the harmonious interplay of data analytics, data privacy, and security fosters a disciplined and risk-aware environment. Stakeholders operate within established boundaries, guided by stringent controls and prudent data practices. As organizations navigate the landscape of a Traditional data culture, they cultivate a foundation of trust, accountability, and reliability in their data-driven pursuits.

Analytics, Privacy, and Security in a Protectionist Data Culture

In the context of a Protectionist data culture, the integration of data analytics, data privacy, and security takes on a vigilant and cautious demeanor, reflective of the values embedded in the "Protectionist" quadrant of the Usage and Flow Data Culture Model. A Protectionist data culture places paramount importance on shielding sensitive information, safeguarding against risks, and upholding a fortress of data integrity.

Data Analytics in a Protectionist Data Culture

In a Protectionist data culture, data analytics are conducted with a strong emphasis on risk assessment and scrutiny. Analytics initiatives are pursued to uncover insights while minimizing exposure to potential threats. Data experts employ advanced analytical techniques, meticulously assessing the implications of each insight before proceeding.

Strategies for Data Analytics in a Protectionist Data Culture

1. Threat Modeling: Integrate threat modeling into the data analytics process, evaluating potential vulnerabilities and risks associated with each analytical endeavor.

2. Secure Data Analytics Environments: Establish isolated environments for data analytics, segregating sensitive data from analysis to minimize the risk of accidental exposure.

3. Adaptive Analytics Frameworks: Develop adaptive analytics frameworks that adjust analytical processes in response to changing threat landscapes, ensuring ongoing protection.

Data Privacy in a Protectionist Data Culture

Privacy is a foundational tenet of a Protectionist data culture, and data privacy measures are characterized by an uncompromising commitment to safeguarding individual rights and confidential information. Data sharing is highly regulated, with stringent controls in place to prevent unauthorized access.

Strategies for Data Privacy in a Protectionist Data Culture

1. Granular Consent Mechanisms: Implement granular consent mechanisms that allow individuals to specify precisely how their data can be used, ensuring explicit control over personal information.

2. Privacy-Preserving Techniques: Leverage privacy-preserving technologies, such as differential privacy, to enable data analysis without compromising individual privacy.

3. Data Anonymization: Apply robust data anonymization techniques to ensure that data used for analysis cannot be linked back to specific individuals.

Data Security in a Protectionist Data Culture

Security in a Protectionist data culture is characterized by fortified defenses, stringent access controls, and a comprehensive approach to risk mitigation. Data security measures are designed to create an impervious barrier against unauthorized access and breaches.

Strategies for Data Security in a Protectionist Data Culture

1. Zero-Trust Architecture: Adopt a zero-trust architecture that verifies and authenticates every data access request, regardless of the source, to minimize potential vulnerabilities.

2. Multi-factor Authentication (MFA): Implement MFA for all data access points, ensuring that only authorized personnel can gain entry to sensitive data resources.

3. Continuous Threat Monitoring: Employ advanced threat monitoring tools and techniques to detect and respond to potential security breaches in real time.

In a Protectionist data culture, harmonizing data analytics, data privacy, and security embodies an unyielding commitment to data fortification and risk mitigation. Stakeholders operate within an ecosystem characterized by unwavering diligence and robust safeguards, ensuring that data assets remain impervious to external threats. As organizations navigate the landscape of a Protectionist data culture, they cultivate a fortress of data resilience, fostering an environment of unwavering trust, vigilance, and data integrity.

Analytics, Privacy, and Security in a Preservationist Data Culture

Within the confines of a Preservationist data culture, the fusion of data analytics, data privacy, and security paints a tableau of careful curation and preservation, echoing the values encapsulated in the "Preservationist" quadrant of the Usage and Flow Data Culture Model. A Preservationist data culture operates with reverence for historical context, cherishing data as a valuable artifact while navigating the nuanced landscape of modern data practices.

Data Analytics in a Preservationist Data Culture

In a Preservationist data culture, data analytics are conducted with a focus on heritage and continuous improvement, extracting insights to illuminate past patterns and trends. Analytics endeavors are anchored in the goal of uncovering insights that inform decisions while retaining data in their original context.

Strategies for Data Analytics in a Preservationist Data Culture

1. Historical Analytics Framework: Develop analytics frameworks prioritizing historical data integrity, allowing for longitudinal analysis without compromising the original data context. The goal of these historical frames is the identification of improvement opportunities to perpetuate the organization.

2. Interdisciplinary Collaboration: Collaborate with historians, archivists, and domain experts to ensure that data analytics align with the Preservationist ethos, capturing the essence of past eras.

3. Data Restoration Techniques: Implement data restoration techniques to revitalize and analyze legacy datasets, uncovering insights while preserving the historical narrative.

Data Privacy in a Preservationist Data Culture

Privacy in a Preservationist data culture is characterized by a delicate balance between historical transparency and contemporary privacy considerations. Data is treated as an archival artifact, with a commitment to protecting individual privacy while respecting the historical significance of the data.

Strategies for Data Privacy in a Preservationist Data Culture

1. Contextual Privacy Frameworks: Develop contextual privacy frameworks that contextualize data privacy within historical contexts, ensuring that privacy measures honor the values of the Preservationist culture.

2. Anonymization and Historical Integrity: Apply anonymization techniques that protect individual privacy while preserving historical data integrity, enabling meaningful analysis without compromising privacy.

3. Ethical Archiving Practices: Employ ethical archiving practices that adhere to modern data privacy principles while honoring the cultural and historical importance of the data.

Data Security in a Preservationist Data Culture

Security in a Preservationist data culture is a guardian of historical integrity, ensuring that data remains unaltered and protected from unauthorized access. Preservationist security measures focus on safeguarding data's authenticity while embracing contemporary security best practices.

Strategies for Data Security in a Preservationist Data Culture

1. Immutable Data Storage: Utilize immutable data storage mechanisms that safeguard the historical integrity of data records, preventing unauthorized alterations.

2. Digital Signature and Authentication: Implement digital signature and authentication protocols that validate the authenticity of historical data while providing a secure foundation against tampering.

3. Cultural Heritage Preservation: Align data security efforts with cultural heritage preservation initiatives, ensuring that security measures harmonize with the reverence for historical artifacts.

In a Preservationist data culture, the convergence of data analytics, data privacy, and security engenders a tapestry of meticulous preservation, historical reverence, and contemporary safeguards. Stakeholders navigate a landscape where data is cherished both as a window to the past and a source of current insights. As Preservationist organizations traverse the delicate balance between history and modernity, they cultivate an environment where data use thrives to maintain the existing heritage and create a solid future of improvement.

Analytics, Privacy, and Security in a Progressive Data Culture

In a Progressive data culture, the fusion of data analytics, data privacy, and security takes on a unique and dynamic character, reflecting an ethos of openness, collaboration, and innovation. Rooted in the "data democratization" edge of the "Data Flow" vector of the UFDCM, a Progressive data culture champions the free flow and its usage for insights and knowledge, empowering stakeholders at all levels to actively engage with data.

In this culture type, data analytics serve as a catalyst for exploration, experimentation, and transformative decision-making. Data is perceived as a strategic asset that fuels innovation and guides organizational strategy. Analytics tools and platforms are readily accessible, allowing employees to conduct exploratory analyses, generate actionable insights, and uncover hidden patterns. A rich ecosystem of visualization tools and self-service analytics facilitates interactive data exploration, enabling users to intuitively interact with data, generate real-time insights, and foster data-driven creativity.

Strategies for Data Analytics in a Progressive Data Culture

1. Self-Service Analytics: Empower employees with user-friendly self-service analytics tools that enable them to explore and analyze data without relying on specialized technical skills.

2. Data Literacy Initiatives: Implement training programs and workshops to enhance data literacy across the organization, ensuring that employees can effectively interpret and leverage data insights.

3. Cross-Functional Collaboration: Foster collaboration between departments by providing shared analytics platforms, promoting the exchange of insights, and encouraging a holistic view of data.

Data Privacy in a Progressive Data Culture

While data democratization is a hallmark of a Progressive data culture, it is essential to establish robust data privacy measures that safeguard individual rights and maintain stakeholder trust. Data privacy is woven into the fabric of data sharing, ensuring that sensitive information is appropriately protected while enabling transparent and ethical data practices.

Strategies for Data Privacy in a Progressive Data Culture

1. Privacy by Design: Integrate privacy considerations from the outset of data initiatives, embedding privacy controls and protections into data workflows.

2. Granular Access Controls: Implement fine-grained access controls that allow data sharing while ensuring that only authorized individuals can access specific datasets.

3. Anonymization and Pseudonymization: Utilize advanced techniques like anonymization and pseudonymization to protect individual identities and sensitive information during data sharing.

Data Security in a Progressive Data Culture

Data security in a Progressive data culture is characterized by a balance between openness and protection. While data flows freely, principled security measures ensure the confidentiality, integrity, and availability of data assets. Security protocols are seamlessly integrated into data pipelines and analytics platforms, mitigating risks and bolstering stakeholder confidence.

Strategies for Data Security in a Progressive Data Culture

1. Multilayered Security: Implement a multilayered security approach that encompasses network security, encryption, access controls, and user authentication.

2. Real-Time Monitoring: Deploy advanced monitoring tools that provide real-time visibility into Data Flows, enabling prompt detection and response to security incidents.

3. Security Awareness Training: Conduct regular security awareness training to educate employees about potential threats, best practices, and their role in maintaining data security.

In a Progressive data culture, the harmonious integration of data analytics, data privacy, and security nurtures a vibrant ecosystem of innovation and collaboration. Stakeholders are empowered to explore data insights, confident that robust privacy and security safeguards underpin their endeavors. As organizations embrace the Progressive data culture, they forge a path toward informed decision-making, value creation, and sustained excellence.

Summary

In the pursuit of fostering a mature and thriving data culture, the convergence of data analytics, data privacy, and security emerges as a pivotal nexus of utmost importance. Throughout this chapter, we have navigated the intricate terrain of data culture through the lens of the Usage and Flow Data Culture Model (UFDCM), unveiling the nuanced interplay among these essential facets.

Our exploration has guided us through the dynamic vectors of the UFDCM, revealing how distinct data culture types span a spectrum of attitudes and practices. The "Progressive data culture" type underscores the empowerment of stakeholders to extract valuable insights from an open and shared data environment, fostering a culture of democratized data access. Conversely, the "Protectionist data culture" type accentuates the critical nature of safeguarding data integrity through controlled and authorized data sharing, underscoring the significance of data security.

Of significance is the "Preservationist data culture" type, which casts light on the foundational role of data privacy in cultivating trust and responsibility within an organization's data landscape. This trust serves as the cornerstone, fortified through the implementation of robust data privacy and security measures. These measures serve as the bedrock not only for the protection of sensitive information but also for the maintenance of regulatory compliance and the enhancement of stakeholder confidence.

In the realm of data culture, data analytics serve as the catalysts propelling informed decision-making, guiding organizations toward strategic triumph. However, the potency of these insights reaches its zenith when nurtured within the fertile ground of a thriving data culture, enabled by the principles of data privacy and security. The symbiotic enrichment between these elements transcends the realm of technical prowess; it forms the very essence upon which a mature data culture flourishes.

As we conclude this chapter, it is imperative to reemphasize the indelible significance of data analytics, data privacy, and security in shaping a data culture that evolves from its nascent stages to full maturity. This culture envisions data not merely as a tool but as a living entity that informs, steers, and catalyzes innovation. From the democratization of data access in Progressive cultures to the fortified ramparts of controlled access in Traditional cultures, and from cultivating a privacy ethos in Preservationist cultures to architecting resilient security frameworks in Protectionist cultures, these elements collectively compose the symphony that orchestrates a thriving data culture. As organizations embrace these tenets, they lay the foundation for the unceasing growth, adaptation, and metamorphosis that characterize the transformative journey of a data culture toward excellence.

CHAPTER 8

Nurturing Communication and Collaboration

Sir Malcolm Bradbury said, "Culture is a way of coping with the world by defining it in detail." Selecting the right culture will offer a detailed definition or focus to achieving these goals. In the Usage and Flow Data Culture Model (UFDCM), each culture has varying degrees of Data Flow and Data Usage to realize the organization's goals. Building a mature data culture goes beyond implementing advanced technologies and tools; it requires communication and collaboration. Communication and collaboration must be focused.

According to Shah (2020), "focus" refers to the ability to direct one's attention and efforts toward a single or set of activities to reach the organization's goals by excelling at those activities. With focus, "collaboration" will have a target to act on. According to Robinson (2022), collaboration refers to the process of two or more individuals or groups working together to achieve a common goal by exchanging ideas and thoughts to realize that goal. Communication fuels the collaboration process. In this context, communication refers to the process of exchanging information and ideas between individuals or groups within an organization (Eads, 2023). With a focus on a particular goal, collaboration can pool the collective ideas to optimize the activity facilitated by communication between the actors.

A mature data culture needs communication and collaboration, and it must be focused. None of the UFDCM cultures prohibits or hinders collaboration or communication, but they channel it through the choices along the Data Flow and Data Usage vectors. This channeling increases focus that enables collaboration and facilitates communication. Selecting the proper culture for the organization is essential to ensure this increased focus.

Communication and collaboration are natural outgrowths of a thriving data culture. When individuals and teams collaborate effectively, they can break down silos, bridge gaps between departments, and leverage diverse perspectives to derive actionable insights from data. Similarly, robust communication channels and platforms facilitate sharing of data insights and practices, ensuring that knowledge is disseminated across the organization and accessible to all relevant stakeholders.

This chapter digs deeper into the various aspects of building a data culture and its ability to nurture communication and collaboration. We begin with an expanded discussion of the interplay between focus, collaboration, and communication. With this interplay explained, we turn to the seven components of a mature data culture to offer levers for leadership to use. Next, we explore focus, collaboration, and communication in the Progressive, Traditional, Preservationist, and Protectionist cultures of the UFDCM. This exploration will include the positives, negatives, and opportunities for leadership of each culture to build focus, increase collaboration, and enhance communication.

Focus, Collaboration, and Communication Interplay

The interplay between focus, collaboration, and communication forms a dynamic and interconnected relationship within organizations. These three concepts are not isolated but rather mutually reinforcing, with each influencing and enhancing the others. When leaders select a data culture, they activate the data to improve focus, encourage collaboration, and enhance communication. The relationships between the trio are at the center of the data culture and decision-making process.

Focus and collaboration share a reciprocal relationship that is pivotal for an organization's progress. A well-defined focus directs the energies and efforts of individuals and teams toward specific goals or tasks. This sense of purpose is the guiding force that ensures everyone is aligned and working cohesively. When teams have a clear focus, they bring a heightened level of commitment and dedication to collaborative efforts. They are more likely to synergize their skills and knowledge to achieve shared objectives. In this way, the focus becomes the driving force behind effective collaboration.

Moreover, collaboration has the remarkable ability to enhance and enrich focus within an organization. Collaborative endeavors naturally involve the integration of diverse perspectives, which can illuminate new angles and insights previously unnoticed. This collective input not only refines the existing focus but also broadens it, allowing for a more comprehensive understanding of the task at hand. Through collaborative brainstorming and problem-solving, teams can fine-tune their focus and recalibrate their efforts to align with the ever-evolving landscape of challenges and opportunities.

At the heart of successful collaboration lies effective communication. Collaboration thrives when team members openly share information, ideas, and feedback. Clear and transparent communication ensures that everyone is on the same page, fostering an environment of mutual understanding and trust. Without effective communication channels, collaboration can falter as misunderstandings arise, tasks are duplicated, or efforts become misaligned. Timely and accurate communication fuels the engine of collaboration, enabling teams to coordinate their actions, make informed decisions, and capitalize on collective strengths.

Conversely, communication is enriched and amplified through collaboration. When individuals from diverse backgrounds come together to collaborate, they bring a wealth of knowledge, experiences, and perspectives. Through active dialogue and exchange, participants can refine their communication skills, adapt their language to accommodate varying viewpoints, and develop a shared vocabulary that enhances overall clarity. Effective collaboration serves as a practice ground for learning how to communicate more inclusively and effectively, which translates into improved communication throughout the organization.

Communication is the backbone of organizational focus, ensuring that all stakeholders are informed about priorities, strategies, and shifts in direction. A steady flow of communication keeps individuals attuned to the organization's evolving needs, helping them realign their efforts, as necessary. Through clear communication channels, leaders can reinforce the importance of certain tasks or goals, creating a collective sense of purpose that bolsters focus. Regular updates and transparent communication provide the necessary context for employees to make informed decisions and effectively channel data and their energies.

Moreover, focus and communication reciprocally enhance each other's effectiveness. When individuals are focused on specific goals or tasks, their communication becomes more purposeful and directed. Clarity of purpose guides the information shared, ensuring it is relevant and aligned with the immediate objectives. On the other side, effective communication enables individuals and teams to share insights, feedback, and progress updates, which in turn aids in maintaining a sharp focus on goals by providing a clear picture of where efforts stand and how they contribute to the larger mission.

The interplay among focus, collaboration, and communication is a dynamic and symbiotic relationship that propels organizational success. Each element reinforces the others, creating a harmonious cycle of achievement. Recognizing and nurturing this interplay is crucial for organizations seeking to excel in today's complex and rapidly evolving landscape. Organizations can foster innovation, adaptability, and resilience by harnessing the power of focus, collaboration, and communication, positioning themselves for sustained growth and excellence.

A Mature Data Culture Enhances Communication and Collaboration

In this section, we explore using the components of a mature data culture as levers to encourage collaboration and enhance communication. The seven components are data strategy, data governance office, data governance program, clearly documented operational and data processes, data analytics, data literacy, and data privacy and security as described in Chapter 2. These components offer leaders powerful tools to mold the culture and increase focus on the organization's goals. As mentioned in the previous section, this increase in focus creates an environment ready for communication and collaboration.

1. Data Strategy: Data strategy fosters collaboration by creating a shared vision and common understanding of how data will be used across departments and teams. When teams are aligned with the same data-driven goals, they are more likely to collaborate and share insights. An effective data strategy requires clear and consistent communication to ensure all stakeholders are on the same page. The data strategy establishes communication channels for sharing insights, progress updates, and outcomes derived from data analyses, promoting a culture of informed decision-making.

2. Data Governance Office (DGO): The DGO fosters collaboration by creating a common framework for data-handling practices. With the goal of more informed decisions, the DGO encourages teams to collaborate to maintain accurate and trustworthy data by defining roles, responsibilities, and ownership. Additionally, the DGO's role in overseeing data quality and integrity fosters a sense of shared ownership, promoting a collaborative culture where teams actively contribute to maintaining and improving data assets.

   Effective communication is essential to the DGO's function for successfully implementing data access and usage policies. The CDO and DGO are central points of contact for communicating data governance guidelines and practices across the organization. By facilitating transparent communication channels about data governance policies, the DGO ensures that teams and individuals understand the rules and standards for data handling. This open communication fosters a culture of shared understanding, where employees are informed about data-related expectations and practices.

3. Data Governance Program (DGP): The data governance program fosters collaboration by creating shared ownership of data and promoting cross-functional engagement. The DGP establishes a collaborative network responsible for data-related activities by designating data stewards, custodians, and users. For example, the modern data catalog centralizes descriptions and inventories to increase collective knowledge. This initiative of the DGP facilitates collaboration by defining data-related processes and workflows, providing common definitions and an understanding of how data flows across teams.

   Effective communication drives the data governance program's success, as it promotes transparent discussions about data-related practices and policies. The DGP establishes communication channels to convey data governance guidelines, standards, and updates, ensuring employees are well informed about data-related expectations. This transparent communication cultivates a shared understanding of data practices and encourages open dialogues about data quality, privacy, and security. The DGP also enhances communication by providing a platform for discussing data-related challenges, sharing best practices, and seeking guidance from data stewards and custodians.

4. Clearly Documented Operational and Data Processes:
    Clearly documented operational and data processes promote collaboration by providing a shared reference point for how data flows and tasks are interdependent. Documented processes encourage teams to work cohesively, leveraging each other's expertise and contributions to achieve common data-related objectives. Furthermore, documented processes serve as a foundation for cross-functional collaboration allowing teams to collaborate on tasks that span different processes and departments, promoting a culture of collective problem-solving and innovation.

    Documented operational and data processes significantly enhance communication within an organization by providing a clear and common understanding of data-related workflows. By creating a shared language and understanding of processes, teams can engage in more effective and meaningful discussions about data-related activities. Documented processes also facilitate transparency in communication, as they provide a platform for discussing potential improvements, identifying challenges, and sharing best practices, fostering an environment of open and informed communication.

5. Data Literacy: Data literacy promotes collaboration by creating a common language and understanding of data concepts among employees. When individuals across different departments and teams share a similar level of data literacy, they can engage in more productive and collaborative discussions. Additionally, data literacy enables employees to contribute meaningfully to cross-functional projects that require data analysis and interpretation, fostering a collaborative culture across the organization.

Data literacy significantly enhances communication within an organization by enabling individuals to communicate data insights clearly and effectively. Visualizations, charts, and summaries become more accessible tools for communicating insights to a wider audience. This enhances communication by making data-driven information more digestible and engaging for both technical and nontechnical stakeholders. Furthermore, data literacy facilitates open discussions about data quality, sources, and methodologies, promoting a transparent and collaborative environment where teams can share insights, ask questions, and explore data-driven opportunities together.

6. Data Analytics: Data analytics encourages collaboration by bringing together cross-functional teams to analyze and interpret data insights. Collaborative efforts are often necessary to extract valuable insights from complex datasets. Cross-functional collaboration ensures that insights are comprehensive and relevant as team members contribute their unique insights and perspectives. Moreover, data analytics projects foster an environment where team members share knowledge, engage in joint problem-solving, and collectively contribute to data-driven initiatives, promoting a culture of collaborative exploration and analysis.

Data analytics significantly enhances communication within an organization by presenting complex data insights through visualizations and clear summaries. Visual representations, such as charts, graphs, and dashboards, facilitate the transmission of intricate concepts and findings. These visual aids make it easier for teams to share data-driven insights effectively, enabling both technical and nontechnical stakeholders to grasp key takeaways. Effective communication of data insights encourages transparency and shared learning, fostering a collaborative culture of data-driven decision-making.

7. Data Privacy and Security: Data privacy and security measures foster collaboration by building trust among teams and departments. When employees know that their data-handling practices adhere to robust security protocols, they are more inclined to collaborate and share information with colleagues across different areas. Collaboration flourishes when data is shared responsibly and securely, enabling teams to leverage each other's insights and expertise. Data privacy and security efforts enhance overall communication within an organization. Clear protocols and transparent communication about security practices ensure that all stakeholders are well informed about data protection measures. Furthermore, data privacy and security measures establish a framework for responsible data sharing, encouraging informed and transparent communication about data-related activities.

Data privacy and security efforts enhance overall communication within an organization. Clear protocols and transparent communication about security practices ensure that all stakeholders are well informed about data protection measures. Furthermore, data privacy and security measures establish a framework for responsible data sharing, encouraging informed and transparent communication about data-related activities.

Communication, Collaboration, and Breaking Down Silos

Silos within organizations have long been recognized as barriers to effective communication, collaboration, and knowledge sharing. What are silos? "In business, organizational silos refer to business divisions that operate independently and avoid sharing information. It also refers to businesses whose departments have siloed system applications, in which information cannot be shared because of system limitations.

The silo mentality is seen as a top-down issue arising from competition between senior managers. The protective attitude toward information begins with management and is passed down to individual employees. It also may be seen between individual employees, who may hoard information for their benefit. It is often found between employees of competing departments, such as marketing and sales, where some assigned duties overlap" (Kenton, 2020).

As described by Kenton (2020), silos differ from leadership limiting data flow for compliance, security, or intellectual property protection. In silos, personal or departmental agendas are at play rather than the organizational good. In the UFDCM Data Flow vector, all required data to support a process will flow even when data flow is limited. The organizational decision to limit data flow is made with the organization's good in mind. Silos cause processes to be underoptimized as required data is blocked improperly.

In the context of building a data culture, silos can have a detrimental impact on the organization's ability to harness the full potential of its data assets. The negative impact of silos on data culture is significant and directly impacts the ability to effectively communicate. When data is confined within silos, it becomes fragmented and lacks a holistic view, limiting its usability and hindering the organization's ability to derive actionable insights. Silos also contribute to duplicate efforts, redundant data collection, and inconsistent data quality, all of which impede the organization's progress in leveraging data effectively.

Overcoming these challenges is never easy. However, the UFDCM provides guidance to overcome these challenges and foster a culture of enhanced communication and cross-functional collaboration. This model emphasizes the importance of breaking down silos and promoting collaboration across departments and teams, recognizing that a collective effort is necessary to fully harness the value of data.

Applying the UFDCM, organizations can implement several strategies to encourage cross-functional collaboration and break down silos. First and foremost, it is essential to establish a shared vision and purpose around data. This involves creating a clear understanding of how data can drive value across the organization and aligning all departments and teams toward a common goal. Second, organizations can facilitate collaboration by creating cross-functional teams or task forces focused on specific data initiatives. These teams should comprise individuals from different departments with diverse skill sets, ensuring a variety of perspectives and expertise. By working together, these teams can break down silos, promote knowledge sharing, and enable the cross-pollination of ideas and insights.

In addition to cross-functional teams, organizations should invest in creating platforms and spaces for collaboration and knowledge exchange. This can include implementing collaborative tools and technologies that enable real-time sharing of data, insights, and best practices. By providing accessible platforms, organizations encourage open communication and collaboration among individuals and teams. Furthermore, the UFDCM emphasizes the need for leadership support and sponsorship to promote cross-functional collaboration. Leaders play a crucial role in fostering a collaborative environment by incentivizing collaboration, breaking down hierarchical barriers, and encouraging open dialogue and idea sharing.

By embracing the principles outlined in the model, organizations can create a culture where open communication and cross-functional collaboration become the norm rather than the exception. Breaking down silos and promoting collaboration enable organizations to tap into the collective intelligence and expertise of their workforce, leading to more innovative solutions, better decision-making, and improved outcomes. By recognizing the negative impact of silos and emphasizing the importance of collaboration, organizations can break down barriers and foster improved communication and collaboration across departments and teams to fully unlock the potential of their data assets.

Communication and Collaboration Between Data Professionals and Business Stakeholders

As mentioned earlier in this book, technology and tools have increased data accessibility. Some stakeholders throughout the business with the knowledge of the tools and data can perform self-service report building or limited analytics. Even with these technologies, advanced analytics may require expertise in statistical modeling, analysis techniques, and specialized tools. The data governance office (DGO) will have data professionals to aid business stakeholders in this advanced analysis. Collaboration and communication between these data professionals and business stakeholders are vital to smoothly facilitating these efforts.

The tools and self-service capabilities available for business stakeholders to access and analyze data have become commonplace in industry. This technology plays a crucial role in fostering collaboration between data professionals and business stakeholders. Utilizing data visualization tools, self-service analytics platforms, and interactive dashboards can enable business stakeholders to explore and interact with data directly, reducing reliance on data professionals for every analysis. This empowers business stakeholders to leverage data in their decision-making processes and encourages a culture of data-driven insights across the organization.

Collaboration between data professionals and business stakeholders drives impactful insights and informed decision-making. Data professionals possess the technical expertise and knowledge to extract and analyze data, while business stakeholders provide the context and domain expertise necessary for deriving actionable insights. By fostering effective collaboration between these two groups, organizations can unlock the full potential of their data assets and drive meaningful business outcomes that are easy to articulate throughout the organization.

The collaboration between data professionals and business stakeholders is crucial because it bridges the gap between technical data expertise and business needs. Data professionals typically possess deep knowledge in data analysis, data modeling, and data visualization techniques. On the other hand, business stakeholders have a deep understanding of the organization's goals, challenges, and industry dynamics. By combining their expertise, data professionals and business stakeholders can derive insights that are not only technically accurate but also relevant and actionable in the organizational context.

However, fostering collaboration between these two groups can present challenges. Communication gaps, differing priorities, and a lack of understanding of each other's roles and responsibilities can hinder effective collaboration. Data professionals may struggle to translate complex technical concepts into business insights, while business stakeholders may have difficulty articulating their data requirements. Overcoming these challenges requires a concerted effort to establish a collaborative environment and promote mutual understanding.

The UFDCM emphasizes the need for collaboration between data professionals and business stakeholders. The model underscores the importance of establishing cross-functional teams and data governance structures that bring together individuals from both technical and business backgrounds. By creating forums for collaboration, such as data governance councils or cross-departmental data teams, organizations can facilitate ongoing dialogue, knowledge sharing, and joint decision-making.

In addition to the model, organizations can implement specific strategies to foster collaboration. For instance, regular meetings or workshops can be organized to enhance mutual understanding of data and business needs. Data professionals can actively engage with business stakeholders to gather requirements and provide insights, ensuring that the analyses are aligned with the organization's goals. Similarly, business stakeholders can provide feedback on the relevance and impact of data-driven insights, further refining the analytical processes.

By utilizing the UFDCM, organizations can bridge the gap between data professionals and business stakeholders. The model promotes collaboration, mutual understanding, and joint ownership of data-driven initiatives. By establishing clear roles, responsibilities, and communication channels, organizations can ensure that data professionals and business stakeholders work together seamlessly to drive data-driven decision-making and foster a strong data culture. By recognizing the importance of their respective expertise, overcoming challenges, and leveraging the Usage and Flow Data Culture Model, organizations can bridge the gap between data and business teams, driving effective collaboration and unlocking the true potential of their data assets.

Communication and Collaboration in the UFDCM

Each culture in the Usage and Flow Data Culture Model (UFDCM) has unique perspectives, goals, and limitations that focus the culture on desired outcomes. In each culture, positive outcomes should be the norm; however, if not effectively managed, negative outcomes inevitably will occur. These negative outcomes include data silos, unfocused collaboration, and disconnected communication. In this section, we cover each culture and its methods for increasing focus, encouraging collaboration, and enhancing communication. As each culture is covered, the positives and potential negatives are discussed. At the end of each cultural analysis, a bulleted list for leaders demonstrates the importance of leadership engagement in establishing and sustaining a mature data culture.

Communication and Collaboration in a Progressive Culture

A Progressive culture is characterized by the free flow of data and the availability of a full suite of data analytic capabilities. The Progressive culture can increase focus by leveraging advanced data insights to guide decision-making. With democratized data flow and comprehensive Data Usage, teams can align their efforts with data-driven strategies that yield measurable results. The wealth of information allows organizations to prioritize initiatives with clarity, channeling resources toward high-impact projects that align with their mission. This focus offers great freedom for communication and collaboration.

Collaborative efforts thrive in a Progressive culture, where a diverse range of data-driven insights encourages cross-functional collaboration. Teams can draw from a rich pool of information to collaborate on innovative projects, leveraging varied perspectives to develop well-rounded solutions. The availability of comprehensive data fosters discussions grounded in evidence, promoting open dialogue and informed problem-solving.

Collaboration with data in a Progressive culture yields a treasure trove of rich insights and innovative solutions. Integrating diverse perspectives and expertise from various functions and domains amplifies the depth of analysis. Cross-functional collaboration enhances understanding of complex trends and correlations, unearthing insights that might not be apparent through isolated efforts. This fusion of knowledge fosters creative problem-solving and the discovery of transformative approaches that drive the organization forward.

Data-driven collaboration in a Progressive culture fosters a comprehensive understanding of organizational performance. By pooling data from diverse sources and functions, teams gain a holistic perspective that transcends silos. This interconnected view empowers stakeholders to make informed decisions considering the broader implications of their choices. The collaborative effort leads to a unified narrative of organizational health and trends, ensuring a well-rounded approach to decision-making.

Collaborating with data accelerates decision-making within a Progressive culture. The collaborative exchange of insights and findings allows teams to quickly analyze complex datasets, share diverse viewpoints, and collectively arrive at well-informed decisions. The iterative feedback loop of collaboration minimizes delays, enabling the organization to be agile and responsive in a rapidly evolving landscape.

Despite its advantages, collaboration with data in a Progressive culture can sometimes introduce complexity. Using advanced analytics techniques and diverse data sources might generate intricate insights that are challenging to communicate effectively, especially to nontechnical stakeholders. Striking a balance between sharing sophisticated insights and ensuring clarity and accessibility becomes a crucial consideration to avoid overwhelming the audience.

The wealth of insights in a Progressive culture can potentially lead to information overload. Collaborators might be inundated with abundant data, making it challenging to discern key takeaways from the noise. Effective communication strategies must be employed to distill and prioritize insights, ensuring stakeholders can focus on critical information without being overwhelmed by the sheer volume of data.

Collaborating using data while promoting diverse perspectives might introduce the challenge of misalignment. Differing interpretations of data and insights could lead to confusion or disagreements among cross-functional teams. Ensuring clear and open communication channels is essential to bridge these gaps and align stakeholders around a unified understanding of data-driven insights.

With advanced data insights at their disposal, communication is enriched in a Progressive culture. Clear and data-backed insights can be effectively communicated through visualizations and reports, enhancing understanding among stakeholders. Transparent discussions about data-driven findings foster a culture of open communication where teams can engage in meaningful conversations about insights and implications.

In a Progressive culture, communicating with data offers the advantage of extracting rich and comprehensive insights. The diversity of data sources and utilization of advanced analytics techniques enable teams to uncover intricate patterns and correlations that might otherwise go unnoticed. Collaborative efforts to analyze these insights from various angles result in a holistic understanding of complex phenomena. The synthesis of data-driven insights empowers stakeholders to make informed decisions grounded in a deep comprehension of the underlying factors.

Collaboration with data in a Progressive culture fosters an environment of innovation. The availability of diverse data sources and the willingness to explore advanced analytics techniques provide teams with the tools to push the boundaries of conventional problem-solving. Collaborators are encouraged to think outside the box and test novel approaches. This dynamic exchange of ideas fuels the generation of innovative solutions that can lead to breakthroughs, driving the organization forward.

Data communication in a Progressive culture has the potential to engage stakeholders through dynamic and interactive means. Complex insights can be conveyed through visually compelling data visualizations and interactive tools. These engaging formats capture the attention of stakeholders, encouraging them to delve deeper into the insights presented. The interactivity fosters exploration and a deeper connection with the data, enabling stakeholders to gain a more profound understanding of the information at hand.

Despite its benefits, communicating with data in a Progressive culture can present challenges due to the complexity of insights. The wealth of data and the application of advanced analytics techniques might lead to intricate visualizations or interpretations that are difficult for nontechnical individuals to grasp. Striking a balance between delivering sophisticated insights and ensuring they are accessible to a diverse audience requires careful consideration of communication strategies.

The abundance of data available in a Progressive culture can potentially lead to information overload. While comprehensive insights are valuable, inundating stakeholders with too much information can result in key takeaways being overlooked or diluted. Communicators must prioritize and distill insights to convey the most relevant and impactful information, ensuring stakeholders can focus on the key messages without being overwhelmed by data volume.

In a Progressive culture, where data communication involves complex insights, there is a risk of misinterpretation or confusion among stakeholders with varying levels of data literacy. Sophisticated visualizations or nuanced analytics might not be universally understood. Clear and effective communication strategies are essential to bridge this gap, ensuring that all stakeholders accurately interpret the insights, regardless of their technical backgrounds.

So what can a leader do in a Progressive culture to nurture communication and collaboration? Consider these recommendations:

- Promote Openness: Encourage a culture of open sharing and idea exchange. Emphasize the value of diverse perspectives and create platforms for team members to voice their opinions and insights.

- Facilitate Data Literacy: Provide training and resources to enhance data literacy across the organization. When team members understand data, they are more likely to engage in meaningful discussions and contribute constructively.

- Leverage Advanced Tools: Invest in advanced collaboration tools and data visualization platforms. These tools can facilitate interactive discussions around complex insights, making data-driven discussions more engaging and understandable.

- Recognize Contributions: Acknowledge and reward collaborative efforts. Recognize individuals who contribute to cross-functional collaboration and innovative problem-solving, reinforcing the importance of teamwork.

Communication and Collaboration in a Traditional Culture

In a Traditional culture, focus is increased by channeling Data Usage toward foundational insights. By extracting basic insights from democratized data, organizations can focus on key areas that align with their objectives. This approach enables teams to prioritize essential initiatives and allocate resources strategically. With data flowing freely and Data Usage focused on measurement and diagnostics, communication and collaboration will focus on process improvement and production throughput.

While Data Usage might be basic, collaboration can still flourish in a Traditional culture. Teams can collaborate on projects that benefit from foundational data insights, leveraging shared understanding to address challenges and opportunities. Collaborative efforts are guided by a common foundation, allowing teams to work cohesively toward shared goals.

In a Traditional culture, collaboration with data is marked by simplicity. Using basic measurements and standardized reports results in straightforward communication that is easily accessible to a broad audience. This simplicity ensures that insights are conveyed in a clear and uncomplicated manner, promoting a shared understanding across the organization.

Data communication within a Traditional culture is characterized by unambiguous messaging. Standardized formats and familiar metrics reduce the likelihood of misinterpretation or confusion among stakeholders. The straightforward presentation of data minimizes the risk of conflicting interpretations and enhances the effectiveness of communication.

Collaborating with data in a Traditional culture maintains consistency in communication. Regularly disseminating straightforward data through standardized formats provides a common reference point for decision-making. This consistency allows stakeholders to track performance trends over time and make informed choices based on reliable and comparable data.

While clarity is a strength, collaboration within a Traditional culture might yield limited insights. Relying primarily on basic measurements could prevent stakeholders from gaining a deeper understanding of complex trends and correlations. The absence of more sophisticated analytics might hinder the organization's ability to uncover nuanced insights that can drive innovation and transformative changes.

Data communication in a Traditional culture might miss opportunities for innovation. The focus on basic measurements might result in a lack of deep analysis, preventing the identification of emerging trends or optimization opportunities. A narrow scope of communication could limit the organization's ability to leverage data for uncovering novel insights.

Collaboration within a Traditional culture might inadvertently lead to data communication silos. The focus on standardized reports and basic measurements could narrow the scope of communication, limiting cross-functional collaboration. The absence of a broader context might hinder sharing insights across departments or functions, preventing stakeholders from accessing diverse perspectives.

Communication is pivotal in a Traditional culture to ensure that basic insights are effectively shared. Transparent discussions about available data insights, limitations, and potential applications promote understanding among team members. Clear communication facilitates informed decision-making, allowing teams to align their actions with the available data.

Communicating with data in a Traditional culture offers the advantage of clarity. The straightforward presentation of data through standardized formats and familiar metrics simplifies understanding for a wide range of stakeholders. This clarity ensures that insights are easily accessible and can be comprehended by individuals across the organization, fostering a shared understanding of performance and progress.

Data communication in a Traditional culture maintains consistency using standardized formats. By employing familiar metrics and presentation styles, stakeholders are provided with a common reference point for interpreting data. This consistency aids in comparisons, trend analysis, and decision-making, as stakeholders can easily track changes and developments over time.

Communication with data in a Traditional culture focuses on essential measurements that provide a foundational understanding of performance. By highlighting these fundamental metrics, stakeholders gain insights into the core aspects of the organization's operations. This basic understanding serves as a starting point for informed decision-making and can drive improvements in targeted areas.

Despite its clarity, communication with data in a Traditional culture might result in shallow insights. Relying solely on basic measurements could limit the depth of analysis and the identification of nuanced trends. The absence of more sophisticated analytics might prevent stakeholders from gaining a comprehensive understanding of complex phenomena, potentially missing valuable insights.

In a Traditional culture, where communication emphasizes standardized metrics, there is a risk of inhibiting innovation. Focusing solely on basic measurements might restrict the exploration of innovative solutions or the identification of new avenues for improvement. A narrow scope of data communication could limit the organization's ability to uncover novel insights and drive transformative changes.

While effective in conveying core metrics, communication in a Traditional culture might be limited to standardized reports and predefined metrics. This narrow scope could prevent stakeholders from exploring beyond established parameters and considering additional dimensions of performance. The absence of a broader context might hinder the identification of emerging trends or opportunities.

So what can a leader do in a Traditional culture to nurture communication and collaboration? Consider these recommendations:

- Encourage Cross-Functional Interaction: Organize cross-functional meetings and workshops to encourage interaction among different departments. These interactions provide opportunities for sharing insights and building a more holistic view of the organization's performance.

- Translate Data: Translate data insights into relatable narratives that resonate with different stakeholders. Leaders can bridge the gap between technical metrics and organizational goals, making data-driven communication more accessible.

- Highlight Impact: Emphasize the impact of data on decision-making and performance. Connect data-driven insights to real-world outcomes to motivate teams to engage in data-driven discussions and seek innovative solutions.

- Share Best Practices: Share success stories of groups that effectively used data for problem-solving. This sharing of best practices can inspire others to embrace data collaboration and adopt similar approaches.

Communication and Collaboration in a Preservationist Culture

Preservationist culture maximizes focus by maintaining control over data flow and leveraging advanced data insights. Limited data flow ensures data accuracy, enabling teams to focus on high-quality insights. The controlled environment directs efforts toward well-defined goals, ensuring that resources are allocated to initiatives that align with organizational priorities. Communication and collaboration will support these well-defined goals with advanced techniques for data analysis.

Collaboration is fostered through controlled data flow and advanced Data Usage in a Preservationist culture. Teams can collaborate confidently on projects that rely on accurate data insights, contributing their expertise to collective problem-solving. The shared understanding of data sources enhances cross-functional collaboration and supports well-informed discussions.

Collaborating with data in a Preservationist culture is characterized by precise and accurate communication. The adherence to controlled boundaries ensures that insights are shared within authorized parameters, minimizing the risk of miscommunication. This precision enhances the clarity of communication, enabling stakeholders to make informed decisions based on accurate insights.

One of the key benefits of collaboration with data in a Preservationist culture is the enhanced security of controlled communication. Sharing insights within well-defined limits promotes data security and compliance, reducing the risk of breaches or unauthorized access. This secure environment protects sensitive information, safeguarding the organization's reputation and integrity.

Data-driven collaboration within a Preservationist culture closely aligns with specific goals and objectives. Teams concentrate their efforts on well-defined challenges, ensuring that collaborative endeavors are targeted and purposeful. This focused problem-solving approach optimizes resources and ensures that collaborative efforts yield tangible results.

Despite its advantages, collaboration within a Preservationist culture might be constrained by limited cross-functionality. The controlled boundaries that define the culture could hinder the exchange of insights and perspectives across different departments or functions. This limitation might prevent stakeholders from accessing diverse viewpoints and impede holistic problem-solving.

Due to the approvals and controlled access that characterize collaboration within a Preservationist culture, communication might be slower in addressing dynamic business needs or urgent decisions. The controlled environment, while enhancing security, might impact the organization's agility in responding promptly to evolving situations.

While precise communication is a strength, collaboration within controlled boundaries might discourage exploring new data sources or adopting alternative approaches. The focus on controlled and targeted communication might inadvertently prevent the organization from embracing new methods or insights that fall outside authorized parameters.

Clear communication is essential in a Preservationist culture to ensure that advanced data insights are effectively disseminated. Teams can engage in transparent discussions about data sources, methodologies, and the implications of insights. Communication about data accuracy and relevance fosters a culture of trust and informed decision-making.

Communicating with data in a Preservationist culture is characterized by precision. Collaborative efforts focus on communicating insights within well-defined boundaries, ensuring the information shared is accurate and relevant. This precision enhances the clarity and effectiveness of communication, allowing stakeholders to make informed decisions based on trustworthy insights.

One of the advantages of communicating with data in a Preservationist culture is the enhanced security of controlled communication. The organization minimizes the risk of unauthorized access or data leaks by sharing sensitive data only with authorized individuals. This secure communication ensures that confidential information remains protected, maintaining the organization's integrity and compliance with data regulations.

Collaboration with data in a Preservationist culture aligns with specific goals and objectives. Communication is tailored to address precise challenges, allowing for focused problem-solving and decision-making. This targeted approach ensures that collaborative efforts are efficient and directed toward achieving specific outcomes, facilitating effective problem resolution.

Despite its benefits, communication with data in a Preservationist culture might be limited by rules for cross-functional interaction. The controlled boundaries that define the culture could restrict the exchange of insights across different departments or functions. This limitation might prevent stakeholders from accessing diverse perspectives and impede holistic problem-solving that benefits from a broader range of expertise.

The controlled access and compliance checks that characterize a Preservationist culture could result in slower communication. The need for approvals and adherence to well-defined parameters might delay the dissemination of insights and impact timely decision-making. Striking a balance between controlled communication and responsiveness to dynamic business needs becomes crucial.

While precision is a strength, communication within controlled boundaries might inadvertently restrict innovation. The focus on controlled and targeted communication could limit stakeholders' ability to share and discuss innovative ideas or alternative approaches. The regulated environment might hinder the exploration of uncharted territories and potential breakthroughs.

So what can a leader do in a Preservationist culture to nurture communication and collaboration? Consider these recommendations:

- Define Clear Goals: Set well-defined goals for communication and collaboration within authorized boundaries. This clarity provides a focus for collaborative efforts and ensures that efforts align with organizational objectives.

- Enable Secure Platforms: Provide secure platforms for data sharing and collaboration. Ensuring data remains within authorized parameters while facilitating communication helps build trust in the process.

- Empower Moderators: Appoint moderators or facilitators who can guide data discussions within the controlled boundaries. These facilitators can ensure that conversations remain focused and relevant to the specified goals.

- Encourage Incremental Progress: Encourage incremental steps toward collaboration. Start with smaller initiatives to build trust and demonstrate the benefits of data-driven collaboration within the controlled environment.

Communication and Collaboration in a Protectionist Culture

In a Protectionist culture, focus is optimized by selecting specific data-related initiatives aligned with core objectives. By channeling resources toward targeted projects, organizations ensure their efforts contribute to essential goals. The controlled environment allows for focused allocation of resources, precise interactions, and specificity in communication.

Limited data flow and basic Data Usage do not hinder collaboration in a Protectionist culture. Teams can collaborate within the confines of approved data sources and initiatives, sharing expertise to address specific challenges. The controlled environment encourages collaboration that is precise and directed toward strategic goals.

One of the primary advantages of collaboration with data in a Protectionist culture is the emphasis on data security. The highly controlled communication environment ensures that data remains secure and prevents unauthorized access or leaks. This emphasis on security safeguards sensitive information, preserving the organization's confidentiality and compliance.

Collaboration within a Protectionist culture yields accurate and validated insights. The strict controls on communication ensure that the information shared is precise and aligned with authorized parameters. Stakeholders can rely on the communicated insights as a trustworthy foundation for decision-making.

Data sharing within a Protectionist culture adheres to clear and well-defined boundaries. The strict guidelines and controlled environment reduce the risk of sharing incorrect or unauthorized information. This clarity ensures stakeholders receive communicated insights within authorized limits and focuses collaboration within those boundaries.

While data security is a strength, collaboration within a Protectionist culture might focus on basic measurements, potentially leading to limited insights. The emphasis on data security could restrict the range of insights shared, preventing stakeholders from gaining a deeper understanding of complex phenomena.

The strict controls and security measures that define collaboration within a Protectionist culture might inadvertently create barriers between departments or functions. These barriers could reduce the exchange of insights and impede cross-functional collaboration, limiting the organization's ability to benefit from diverse perspectives. Leaders should pay special attention to the requirements for these controls.

The self-imposed limitations on the Data Flow and Data Usage vectors within a Protectionist culture could restrict stakeholders from exploring insights that could lead to innovation or process improvement. While data security is a priority, the rigid parameters might discourage the exploration of innovative ideas or alternative approaches, potentially hindering the organization's adaptability and growth.

Clear communication is crucial in a Protectionist culture to ensure that teams are aware of accessible data resources and their applications. Transparent discussions about data limitations and potential value promote understanding among team members. Effective communication supports efficient decision-making, even within a limited data landscape.

One of the primary advantages of communicating with data in a Protectionist culture is the emphasis on data security. Highly controlled communication ensures that data remains secure and minimizes the risk of unauthorized access or data leaks. This secure communication safeguards the organization's sensitive information and maintains its integrity.

The strict controls on communication within a Protectionist culture contribute to accurate and validated information being shared. Stakeholders can rely on the precise and controlled nature of communication to ensure that the insights they receive are trustworthy and aligned with the organization's objectives.

Communication within a Protectionist culture is directed toward specific goals and objectives. The stringent controls ensure that insights are shared with a clear purpose, enabling stakeholders to make precise and informed decisions that align with authorized parameters. This focused approach streamlines decision-making processes.

Communication with data in a Protectionist culture might focus on basic measurements, potentially missing deeper insights. The strict controls and emphasis on data security could limit the shared insights, preventing stakeholders from gaining a more comprehensive understanding of complex phenomena. The decision of leaders to reduce the use of analytics further reduces the opportunity for deeper insights.

The strict controls that underpin communication within a Protectionist culture could inadvertently create barriers between departments or functions. The security measures might restrict collaboration and information sharing, potentially leading to isolated silos of knowledge and limiting the organization's ability to benefit from cross-functional insights. Leaders must ensure these controls are "right-sized" to reduce these limitations.

While data security is a priority, the highly controlled communication approach might prevent stakeholders from discussing potential innovations or considering alternative perspectives. The rigid parameters could limit the exploration of new ideas and potentially lead to missed opportunities for growth and improvement. When evaluating processes and interactions, leaders must consider ways such as special projects to ensure avenues for needed innovation or step-wise improvement to exist.

So what can a leader do in a Protectionist culture to nurture communication and collaboration? Consider these recommendations:

- Create Communication Frameworks: Develop well-defined communication frameworks that adhere to security protocols. Establish guidelines for sharing data and insights, ensuring communication remains within authorized parameters.

- Provide Authorized Access: Grant authorized access to relevant stakeholders to participate in data discussions. Empower team members to contribute insights while maintaining the necessary controls.

- Facilitate Secure Discussions: Implement secure communication channels that allow teams to discuss data insights within the protected environment. These channels enable collaboration while ensuring data remains secure.

- By tailoring their approach to the specific culture, leaders can effectively nurture communication and collaboration, fostering a data-driven environment that aligns with the organization's values and goals.

In each culture defined by the UFDCM, organizations can tailor their strategies to increase focus, encourage collaboration, and enhance communication. The UFDCM cultures have a distinctive way of increasing focus and unique positives and negatives that must be managed. For these reasons, leadership plays a central role in setting the proper tone, demonstrating the desired behaviors, and recognizing opportunities to break down barriers to enhance focus and nurture communication and collaboration. By understanding the unique characteristics of each culture, organizations can optimize their data-driven efforts, fostering an environment that aligns with their goals and priorities while cultivating effective focus, collaboration, and communication.

Summary

This chapter explores data culture's ability to increase focus and nurture communication and collaboration. The interplay between this trio is mutually reinforcing, with each enhancing the others. Focus brings clarity to goals and purpose. With this clarity, collaboration between diverse actors in the ecosystem continues to hone solutions. Communication facilitates these interactions, sharpening the focus further.

The seven components of a mature data culture contain tools for increasing focus and tactically increasing communication and collaboration. Data strategy, the data governance office, data governance programs, and clearly documented operational and data processes function as communication vehicles for plans, policies, procedures, and references for Data Usage, quality, and availability. These tools offer the user community a common framework for communication and collaboration. Data literacy and data privacy/security measures offer education and guidelines to ensure communication and data sharing during collaboration are responsibly managed. Finally, data analytics provide tools to explore data cooperatively and create outcomes during collaboration events.

Each data culture has unique perspectives for focus, collaboration, and communication. In all cultures, focused effort must be directly applied to break silo barriers that restrict communication and collaboration. It is never easy, but it is necessary to ensure that data professionals and business stakeholders work together. The cultures seek to encourage focus to provide the proper level of security, privacy, and compliance while encouraging data sharing, exploration, and throughput. Balancing these factors depends on the data culture selected being directly tied to the organization's mission. Leadership must ensure the data culture matches the needs of the organization.

CHAPTER 9

Measuring Success and Sustaining the Data Culture

The Chinese philosopher Lao Tzu said, "The journey of a thousand miles begins with one step." He wasn't wrong because you won't get anywhere unless you start the journey. In the journey toward building a successful data culture, organizations must recognize the significance of measuring success and implementing strategies to sustain the data culture transformation. The Usage and Flow Data Culture Model offers a framework of the different data culture types: Protectionist, Preservationist, Traditional, and Progressive. Each type requires unique approaches for measuring progress and fostering a data-driven environment that aligns with the organization's objectives and long-term vision. This chapter explores the importance of establishing metrics, measuring success, and developing sustainable data culture practices tailored to each quadrant of the UFDCM.

The importance of establishing metrics and indicators can't be overstated. If you want to know your progress for developing a mature data culture, then you must measure it to see how far you've come. In a Protectionist data culture, metrics should focus on data literacy improvement, as employees may lack the skills to confidently engage with data. Some suggested indicators might include tracking the number of employees who have participated in data training programs, the completion rates, and the frequency of data-related queries from employees. In a Preservationist data culture, the focus would be different. The metrics should assess how effectively data

© Gary W. Griffin, David Holcomb 2023
G. W. Griffin and D. Holcomb, *Building a Data Culture*, https://doi.org/10.1007/978-1-4842-9966-1_9

is being utilized to support decision-making. Some indicators that may be useful for this culture type include measuring the increase in data-backed initiatives, the adoption rate of data-driven decision-making, and the level of data integration across departments. In a Traditional data culture, the metrics should focus on enhancing data-driven decision-making and promoting data literacy organization-wide. An example of the indicators for this culture type are assessing the impact of data training programs on decision outcomes, tracking the number of data-informed decisions, and monitoring improvements in data literacy among Traditional data culture employees. For a Progressive data culture, the focus should be on metrics that measure continuous improvement and innovation. Here, you should measure the success of innovative data projects, monitor employee engagement in data exploration, and track the rate of experimentation with new data technologies and tools.

Collect feedback from employees at all levels to understand their experience with the data culture initiative. Use the feedback from employees to identify strengths and weaknesses in the data culture strategy. Analyze data-driven outcomes and decision-making processes to identify areas for improvement and refine the data culture road map as often as needed. Organizational shifts, industry trends, and emerging challenges should be monitored regularly. Align the data culture strategy with evolving business goals and ensure it remains relevant and effective. In all data culture quadrants, strong leadership and advocacy are essential for sustaining the data-driven transformation as described in Chapter 4. Leadership, especially the CDO, should encourage data culture champions and ensure continuity of leadership support. Continuous learning initiatives are vital for sustaining a data culture. Organizations should invest in data training programs to ensure employees remain up to date with the latest data skills and knowledge as enumerated in Chapter 6 on data literacy. Promote cross-functional data utilization and encourage knowledge exchange. Create an environment where you recognize and reward employees who actively embrace data-driven practices and contribute to data-backed initiatives. One possible idea is to incentivize data literacy and data-centric decision-making throughout the organization.

Measuring success and sustaining a data culture are necessary for a successful data-driven organization. By establishing relevant metrics and indicators, continuously refining the data culture strategy, and implementing sustainable practices tailored to each data culture quadrant, organizations can foster a data-driven mindset and achieve data culture excellence. Leaders play a pivotal role in driving this transformation, ensuring that data literacy, data-driven decision-making, and collaboration become ingrained within the organization. Through a commitment to continuous evaluation, adaptation, and the cultivation of a data culture that aligns with its core values, an organization can thrive in a rapidly evolving data-driven landscape and achieve enduring success.

Establishing Metrics and Indicators

In the one thousand mile journey of establishing a mature data culture, the initial steps lay the groundwork for success. The journey begins with the careful selection of the target data culture, followed by a thorough assessment of the current organizational landscape. These initial steps serve as the compass that guides organizations toward embracing data as a transformative force.

As the foundation is laid, the seven key components of a mature data culture (see Chapter 2) come into play, each vital for cultivating a thriving data-driven environment:

1. Data Strategy Development: The formulation and publication of a comprehensive data strategy are essential. This strategic road map outlines how data will be harnessed to achieve organizational goals, setting the tone for a culture where data-driven decisions become second nature.

2. Data Governance Office Establishment: Initiating the data governance office and aligning it with the first budget cycle are critical milestones. This dedicated hub becomes the nucleus for driving data initiatives, ensuring resources are allocated effectively, and fostering a culture of collaboration and innovation.

3. Data Governance Program: Within the data governance program, establishing a modern data catalog becomes a cornerstone. The completeness of data definitions, ownership of data sets, inventories of reports, and user-friendly accessibility define the robustness of data governance efforts. Metrics, such as the catalog's usage and its inclusion in onboarding, offer insights into its impact.

4. Process Management Integration: The integration of process management into data culture establishment involves completing process maps and determining process owners. This integration synergizes data ownership with process ownership, fostering accountability and transparency. The percentage of process maps completed may be one indicator of progress.

5. Data Literacy Nurturing: Comprehensive coverage on data literacy speaks to its important role in a mature data culture. Equipping employees with data skills and fostering a culture of curiosity and exploration accelerate the journey toward data-driven decision-making.

6. Data Analytics Empowerment: Creating conducive environments for data analytics, measuring usage tied to decision-making, and curating an inventory of current data assets form the crux of this component. By linking reports and analytics to corporate performance, the organization creates a direct line between data initiatives and strategic outcomes.

7. Security and Privacy Vigilance: Setting specific, measurable, achievable, relevant, and time-bound (SMART) goals for security, privacy, and compliance awareness at the initiation stage, coupled with ongoing training and validation, ensures that data security and privacy are at the forefront of consideration in its importance. This component safeguards the integrity of the data culture and the organization at large.

As organizations traverse these components, it's essential to align with the overarching mission of cultivating a data culture. The steps taken during establishment serve as the guiding lights, illuminating the path toward a future where data doesn't just inform decisions—it shapes the very fabric of success. In this journey, the establishment of metrics and indicators for each data culture type becomes a compass that ensures progress, sustainability, and the realization of the transformative power of data.

To effectively measure the success of a data culture initiative and ensure its sustainability, organizations must establish key performance indicators (KPIs) that align with the characteristics of each data culture type. Each quadrant of the UFDCM requires specific KPIs to assess progress, track improvements, and drive data-driven decision-making. Let's explore the appropriate KPIs for each data culture type:

1. Protectionist Data Culture Metrics

    a. Data Literacy Rate: Measure the percentage of employees who have participated in data training programs or workshops. This metric reflects the organization's efforts in improving data literacy within this quadrant.

    b. Data Queries: Track the number of data-related queries received from employees. An increase in queries over time indicates growing interest in and engagement with data.

    c. Data Utilization: Assess the frequency of Data Usage in day-to-day tasks among employees. Higher data utilization rates demonstrate a shift toward embracing data-driven practices.

2. Preservationist Data Culture Metrics

    a. Data-Backed Initiatives: Measure the number of projects or initiatives within the Preservationist data culture that are based on data-driven insights. An increase in data-backed initiatives shows a strong integration of data in decision-making.

    b. Data Integration: Evaluate the level of data integration across departments in the Preservationist quadrant. A higher integration score indicates successful data sharing and collaboration.

    c. Data Impact: Assess the impact of data-driven decisions within the Preservationist data culture. This metric indicates how data is influencing business outcomes and strategy.

3. Traditional Data Culture Metrics

    a. Data-Informed Decisions: Monitor the percentage of decisions made within the Traditional data culture that are backed by data. An upward trend in data-informed decisions signifies progress toward a more data-driven environment.

    b. Data Training Impact: Measure the effectiveness of data training programs on employees. Track improvements in data literacy and data skills.

   c. Data Sharing: Evaluate the level of data sharing and knowledge exchange across departments in the Traditional quadrant. A higher data-sharing score indicates enhanced collaboration.

4. Progressive Data Culture Metrics

   a. Data Experimentation Rate: Assess the frequency of data experimentation and exploration among employees. This KPI demonstrates the willingness to explore innovative data practices.

   b. Data Innovation Projects: Track the number of innovative data projects undertaken by Progressive data culture teams. A higher number of data innovation projects shows a culture of innovation.

   c. Data Technology Adoption: Measure the rate of adoption of new data technologies and tools within the Progressive data culture. This KPI reflects the organization's adaptability to technological advancements.

Identifying and implementing relevant KPIs for each data culture type is essential for measuring the success of a data culture initiative and ensuring its sustainability. The ones suggested in this chapter offer a few ideas but by no means are exhaustive. By tailoring KPIs to the unique characteristics of each quadrant, organizations can effectively track progress, drive data literacy, encourage data-driven decision-making, and foster a data culture that aligns with their business objectives. Regularly analyzing KPIs and making data-driven decisions based on the insights obtained will enable organizations to thrive in a data-driven landscape and achieve long-term success.

Aligning Metrics with Organizational Objectives

To effectively measure the success of a data culture initiative and ensure its sustainability, organizations must establish robust data collection and analysis strategies. These strategies play a key role in aligning metrics with the organization's objectives and providing valuable insights for continuous improvement. By leveraging data analytics tools, implementing feedback loops and surveys, and conducting data culture maturity assessments, organizations can gain a comprehensive understanding of their data culture's progress and make data-driven decisions to drive that success.

Data analytics tools are essential for collecting, processing, and analyzing data to derive meaningful insights. By leveraging these tools, organizations can track and measure various data culture metrics efficiently. For instance, clearly define the KPIs and metrics that align with the organization's objectives and data culture type. These metrics should be specific, measurable, achievable, relevant, and time-bound (SMART). Integrate data from various sources, such as employee training records, project management tools, and customer feedback systems. Centralized data integration provides a holistic view of the data culture initiative's impact. The use of data visualization techniques should be utilized to present complex data in a clear and concise manner. Tools like interactive dashboards and well-designed reports help stakeholders to easily interpret and comprehend the data. Set up real-time monitoring systems to track data culture metrics continuously, and get real-time insights that enable prompt decision-making and immediate action.

Feedback loops and surveys are powerful tools to collect insights from employees at all levels of the organization. They provide valuable information about the effectiveness of the data culture initiative and employees' perceptions of the data culture. Some considerations for implementing feedback loops and surveys should include some or all these items. Collection efforts should encourage honest and candid feedback by conducting anonymous surveys. This allows employees to express their opinions without fear of repercussions. It should include regularly scheduled check-ins to obtain ongoing feedback throughout the data culture initiative. This type of feedback helps identify emerging challenges and address them promptly. These types of data collection efforts should include a mix of qualitative and quantitative questions in surveys. This allows for a comprehensive understanding of employees' experiences and perceptions.

Conducting data culture maturity assessments provides a structured approach to evaluating the organization's data culture at various stages of the initiative. These assessments help identify strengths, weaknesses, and areas for improvement. When preparing to do a data culture maturity assessment, consider following these steps when conducting the assessment. First, develop clear criteria to assess the maturity of the data culture in different areas, such as data literacy, data usage, and decision-making processes. Second, define different stages of data culture maturity (e.g., initial, developing, advanced) and evaluate the organization's progress against these stages. Third, involve stakeholders from various departments and levels in the assessment process to gather diverse perspectives. Finally, perform regular assessments at defined intervals to track progress and adapt the data culture strategy accordingly.

Aligning metrics with organizational objectives is essential for measuring the success and sustainability of a data culture. By leveraging data analytics tools, implementing feedback loops and surveys, and conducting data culture maturity assessments, organizations can gain valuable insights into their data culture's progress and make informed decisions to drive continuous improvement. These data collection and analysis strategies empower organizations to optimize their data culture, foster data-driven decision-making, and achieve long-term success in the data-driven landscape.

Measuring Progress and Success

One of the key aspects of measuring the success of a data culture initiative is to track the development of data literacy and skills among employees. As discussed in Chapter 6, data literacy is the foundation upon which a data-driven culture is built, and monitoring the improvement in employee competencies is essential for ensuring the initiative's effectiveness. To achieve this, organizations can implement assessments to evaluate employee competencies and continuously monitor the effectiveness of data training programs.

Assessing employee competencies is a crucial step in measuring the progress of data literacy and skills development within the organization. By conducting assessments at various intervals, organizations can gauge the level of data literacy among employees and identify areas that require further improvement. Here are some considerations for assessing employee competencies. Conduct a pre-training assessment that allows organizations to establish a baseline of employees' data literacy levels. This baseline data serves as a reference point for evaluating the impact of the training programs. Design assessments that target specific data-related skills relevant to different roles within the organization. This tailored approach ensures that employees receive training that is aligned with their job requirements. Regular competency assessments should be used to track employees' progress over time. The continuous monitoring enables organizations to identify trends, address skill gaps, and adapt training strategies as needed.

To ensure the effectiveness of data training programs, organizations must monitor and evaluate the impact of these initiatives. Monitoring data training program effectiveness helps determine if the programs are meeting their intended goals and driving the development of data literacy and skills. You should define clear learning outcomes and objectives for each data training program and evaluate whether employees achieve these outcomes after completing the training. Be sure to gather feedback from employees who have participated in data training programs through post-training surveys. This type of feedback helps to assess the relevance, quality, and practicality of the training content based on the employees' feedback. Analyze employees' performance improvements in their respective roles after participating in data training programs. It's a good idea to look for correlations between training participation and enhanced job performance. Assess the long-term impact of data training programs on the organization's overall data culture, as some of these may not be apparent immediately. Measure how the improvement in data literacy and skills translates into better decision-making and enhanced business outcomes.

Measuring the progress and success of a data culture initiative requires a comprehensive approach to tracking data literacy and skills development among employees. By conducting assessments to evaluate employee competencies and continuously monitoring the effectiveness of data training programs, organizations can gain valuable insights into the impact of their data culture initiative. These measurements provide critical feedback to refine and optimize the initiative over time, ensuring that the organization fosters a sustainable data-driven culture. As employees' data literacy and skills improve, the organization's ability to make data-driven decisions and drive innovation will strengthen, ultimately leading to enhanced business performance and success in a data-driven world.

Evaluating Data-Driven Decision-Making

Data-driven decision-making is at the core of a successful data culture. To measure the success and effectiveness of data-driven initiatives, organizations need to evaluate the outcomes and impact of decisions that are based on data. There are two key aspects of evaluating data-driven decision-making: analyzing data-backed initiatives and measuring decision outcomes and impact.

Analyzing data-backed initiatives involves assessing the process and results of decision-making that relies on data insights. This evaluation helps organizations understand how well data-driven decisions align with organizational objectives and how effectively data is being utilized to drive initiatives. Analyzing data-backed initiatives should involve a number of activities. First, examine how data is incorporated into the decision-making process. Assess whether decision-makers are using data to identify problems, frame questions, analyze information, and make informed choices. Second, ensure that the data used in decision-making is accurate, reliable, and relevant. Evaluate the quality of data sources and the validity of data analysis methods to ensure that decisions are based on reliable information. Third, compare the outcomes of data-driven decisions with those made without data support. Identify instances where data-backed decisions resulted in more favorable outcomes or led to innovative solutions. Fourth, embrace a culture that learns from both successful and unsuccessful data-driven initiatives. Analyze instances where data-driven decisions did not yield the expected results and identify areas for improvement.

Measuring decision outcomes and impact involves assessing the tangible results of data-driven decisions on the organization's performance and success. This evaluation provides insights into the effectiveness of data-driven decision-making in driving positive change and achieving business goals. Measuring decision outcomes and impact must start with defining clear performance metrics and key performance indicators (KPIs) aligned with the objectives of data-driven initiatives. These metrics should be measurable and quantifiable to track progress and success. Monitor how data-driven decisions influence business performance, such as revenue growth, cost savings, customer satisfaction, and operational efficiency. Compare performance before and after data-driven initiatives to assess their impact. Gather feedback from key stakeholders, including employees, customers, and partners, to understand their perceptions of the impact of data-driven decisions on their experiences and interactions with the organization. Assess the long-term impact of data-driven decision-making on the organization's overall success and sustainability. Measure how data-driven initiatives contribute to the organization's strategic goals and competitive advantage.

Evaluating data-driven decision-making is essential for measuring the success of a data culture initiative and ensuring its sustainability. By analyzing data-backed initiatives and measuring decision outcomes and impact, organizations can gain valuable insights into the effectiveness of their data-driven practices. This evaluation allows organizations to identify strengths and areas for improvement in their data culture, refine their decision-making processes, and align data-driven initiatives with organizational objectives. By continuously evaluating data-driven decision-making, organizations can foster a culture of continuous improvement and innovation, ultimately leading to sustained success in an increasingly data-driven world.

Assessing Data Culture Adoption

Assessing data culture adoption is the foundation to understanding how well the data culture initiative is being embraced and integrated within the organization. It involves evaluating how employees engage with data initiatives and the extent of cultural transformation toward a data-driven mindset. Data culture adoption involves determining the level of employee engagement with data initiatives and using cultural transformation assessments.

Employee engagement with data initiatives provides insights into how well employees are participating in and contributing to data-driven projects and activities. Engaged employees are more likely to proactively seek data-driven solutions and apply data insights to their decision-making. Some strategies for assessing employee engagement with data initiatives should involve these areas:

- Measure the level of participation in data-related training programs, workshops, and data-driven projects.

- Track the number of employees involved in data initiatives to gauge the overall interest in data-driven practices.

- Gather feedback from employees on their experience with data initiatives.

- Conduct surveys and focus groups to understand their level of enthusiasm, perceived value, and challenges faced when working with data.

- Evaluate the quality and quantity of contributions made by employees in data-driven projects.

- Assess whether employees actively use data to support their ideas and decision-making.

- Monitor whether employees who actively engage with data initiatives are recognized and rewarded for their efforts. Acknowledging data-driven achievements can encourage further engagement.

Cultural transformation assessment focuses on evaluating the organization's progress in transforming its culture to become data-driven. It involves examining the shifts in attitudes, behaviors, and practices that align with a data-driven mindset. Some strategies for conducting a cultural transformation assessment may include cultural surveys, leadership role modeling, cross-functional collaboration, and cultural values alignment. You should start by administering surveys that capture perceptions and beliefs related to the data culture. Attempt to measure employees' understanding of the importance of data, the level of data transparency, and the extent of collaboration around data use. Evaluate the role of leadership in promoting and exemplifying data-driven practices and assess whether leaders actively seek data insights and make decisions based on data. Analyze the level of cross-functional collaboration related to data initiatives. Measure the extent to which different departments share data, collaborate on data projects, and jointly use data insights. Assess whether the organization's core values align with data-driven principles. Evaluate whether data-driven decision-making is ingrained in the organization's mission and vision.

Assessing data culture adoption is essential for understanding the organization's progress in embracing a data-driven culture. By evaluating employee engagement with data initiatives and conducting a cultural transformation assessment, organizations can gain valuable insights into the effectiveness of their data culture initiatives. This assessment helps identify areas for improvement, address challenges in cultural transformation, and reinforce a data-driven mindset throughout the organization. By actively assessing data culture adoption, organizations can ensure the sustainability of their data culture, foster continuous improvement, and drive long-term success in a data-driven world.

Continuously Refining the Data Culture Strategy

A data culture initiative is not a one-time endeavor but a continuous journey of improvement and evolution. To ensure the sustained success of the data culture, it is essential to continuously refine the data culture strategy based on feedback, learnings from data culture initiatives, and a regular review of the data culture road map. This section discusses the key aspects of continuously refining the data culture strategy by utilizing feedback mechanisms, reviewing the data culture road map, and learning from data culture initiatives.

Feedback mechanisms play a critical role in understanding how the data culture initiative is perceived and experienced by employees and stakeholders. Gathering feedback provides valuable insights into the effectiveness of the strategy and helps identify areas for improvement. Two primary sources of feedback are from employees and data champions/advocates. Regularly seek feedback from employees at all levels of the organization regarding their experience with the data culture initiative. Surveys, focus groups, and one-on-one discussions can be used to understand their perceptions, challenges, and suggestions for enhancing the data culture. Analyze the feedback to identify recurring themes and trends that can guide future improvements.

Data champions and advocates are key drivers of the data culture initiative. Their experiences and observations can offer valuable insights into the impact of the initiative and its alignment with organizational goals. Engage in open discussions with data champions and advocates to gain a deeper understanding of their successes, challenges, and recommendations.

The data culture road map serves as a guiding document that outlines the goals, objectives, and strategies for the data culture initiative. Regularly reviewing the road map ensures that the initiative stays aligned with the organization's changing needs and priorities. The review process should include evaluating whether the goals and objectives of the data culture initiative remain relevant and aligned with the organization's overall strategy. Consider any shifts in business priorities or emerging data-related challenges that may necessitate adjustments to the goals. Assess the progress made in achieving the data culture goals and identify areas where improvements are needed. Determine whether the current strategies are effective and explore new approaches that can enhance the data culture initiative's impact.

Data culture initiatives provide valuable opportunities for learning and growth. By analyzing the outcomes of various initiatives, organizations can gain insights that inform future actions. Data culture initiatives provide an excellent opportunity for learning. Identify data culture initiatives that have been particularly successful in driving positive outcomes. Understand the factors that contributed to their success and consider replicating these practices in other parts of the organization. Data culture initiatives may encounter challenges and setbacks. Embrace these as learning opportunities and use them to identify areas that require improvement. Analyze the root causes of any failures and develop strategies to overcome them in the future.

Continuously refining the data culture strategy is essential for sustaining a thriving data culture within the organization. By utilizing feedback mechanisms from employees, data champions, and advocates, organizations can gain valuable insights into the effectiveness of the data culture initiative. Regularly reviewing the data culture road map ensures that the initiative remains aligned with organizational goals. Learning from data culture initiatives, both successes and challenges, provides valuable knowledge for driving continuous improvement. By embracing a culture of learning and adaptation, organizations can ensure the ongoing success and sustainability of their data culture, fostering a data-driven environment that drives innovation, efficiency, and business growth.

Adapting to Organizational Changes

As an organization evolves, it is crucial to ensure that the data culture initiative remains aligned with the changing business goals and strategic direction. The importance of adapting the data culture to organizational changes and integrating it into strategic planning is examined. Industries are constantly evolving due to technological advancements, market trends, and changes in consumer behavior. To sustain a successful data culture, organizations must stay attuned to industry shifts and understand how these changes impact their data needs and priorities. Regularly assess the data culture's alignment with the evolving demands of the industry and make necessary adjustments to ensure relevance and competitiveness. Organizational changes, such as mergers, acquisitions, restructuring, or expansions, can significantly impact the data culture. During periods of change, it is essential to communicate the importance of data culture and reinforce its integration into the fabric of the organization. Address any challenges or resistance that may arise and emphasize how data-driven decision-making can support successful transitions.

A data-driven culture must be ingrained in the organization's decision-making processes. This requires a proactive effort to ensure that data and analytics are consulted and leveraged at all levels of decision-making. Leaders should emphasize the value of data in informing choices and set the expectation that decisions should be supported by data insights whenever possible. Data culture should play a central role in the formulation of the organization's strategic plans. As part of strategic planning, identify key data-driven objectives and initiatives that align with the overall business strategy. These objectives should be measurable and supported by relevant data and analytics. Data-driven strategy formulation ensures that the organization leverages data to seize opportunities and address challenges effectively.

Adapting the data culture to organizational changes and integrating it into strategic planning are vital for sustaining a thriving data-driven environment. For example, the organization may be a new acquisition and the acquirer may have a different data culture than the acquired. Understanding these dynamics will increase successful establishment of the target data culture. By aligning the data culture with evolving business goals and industry shifts, organizations can remain agile and responsive in an ever-changing landscape. Integrating data culture into strategic planning ensures that data-driven decision-making becomes an inherent part of the organization's DNA.

By embracing data-driven strategies, organizations can capitalize on data insights to drive growth, innovation, and competitive advantage, thereby fostering a data culture that stands the test of time.

Strategies for Sustaining the Data Culture

Sustaining a data culture requires deliberate and continuous efforts to embed data-driven practices into the fabric of the organization. This section outlines key strategies to ensure the long-term success of the data culture initiative. Identifying and empowering data culture champions within the organization can significantly impact the adoption and sustainability of the data culture. These champions act as advocates for data-driven practices, promoting the value of data literacy and inspiring others to embrace data in their decision-making. Encourage collaboration among data champions to share best practices and success stories, fostering a sense of community around data initiatives. Leadership support is critical in driving the data culture forward. To sustain the data culture, ensure that leadership commitment is ongoing and not limited to a short-term initiative. Continuity of support from top-level executives fosters a culture of data-driven decision-making throughout the organization. Regularly communicate the importance of the data culture to leaders, reinforcing its strategic relevance and impact on organizational success.

Data skills and technologies are continually evolving, and organizations must invest in ongoing data training and development initiatives. Offer a variety of learning opportunities, such as workshops, webinars, online courses, and hands-on projects, to cater to different learning preferences and skill levels. Encourage employees to take ownership of their learning journey and provide resources to support their continuous growth in data literacy. Beyond technical skills, nurturing a data-centric mindset is essential for sustaining the data culture. Promote a culture that values data-driven insights and encourages curiosity and critical thinking. Help employees develop an understanding of how data can inform decision-making, improve processes, and drive innovation. Reinforce the message that data is a strategic asset and a valuable resource for achieving organizational goals.

To sustain the data culture, break down silos and promote cross-functional data utilization. Encourage departments to share data and insights, fostering a collaborative approach to problem-solving. Create platforms and forums for knowledge sharing and data exchange, facilitating a culture where data is seen as a shared resource that benefits the entire organization. Facilitate knowledge exchange among employees, data experts, and data novices. Data experts can mentor and guide others, while data novices can offer fresh perspectives and ideas. Establish a culture where sharing data-related experiences and lessons learned is encouraged. Celebrate successful data-driven projects and the collective efforts of teams in adopting data-driven practices.

Provide incentives for employees who actively engage in data training and development, demonstrating their commitment to data literacy. Recognize individuals or teams that achieve notable milestones in their data-driven journey. Incentives can include career development opportunities, performance-based rewards, or public acknowledgment of their data literacy achievements. For example, certificates of achievement or badges that can be used internally or on social media may be provided to employees. Recognize and reward data-driven contributions to projects, initiatives, and decision-making processes. Highlight how data insights have positively impacted outcomes and influenced successful business strategies. By acknowledging and celebrating data-backed contributions, the organization reinforces the value of data-driven behaviors and encourages others to follow suit.

Sustaining a data culture is an ongoing journey that requires dedication, leadership commitment, and continuous learning. By empowering data culture champions, providing ongoing data training and development, fostering data sharing and collaboration, and recognizing data-driven behaviors, organizations can create a sustainable data culture that drives innovation, efficiency, and informed decision-making across all levels of the organization. The strategies outlined in this section contribute to creating a data-driven culture that not only endures but thrives, positioning the organization for long-term success in a data-centric world.

Summary

In this chapter, we explored the critical aspects of measuring success and sustaining a data culture within organizations. We began by establishing the importance of data culture and its impact on decision-making, innovation, and overall business performance. The UFDCM provided valuable insights into the four data culture types highlighting the diverse challenges and opportunities in each quadrant.

Measuring the success of a data culture initiative goes beyond mere data literacy metrics; it involves assessing the organization's overall data culture maturity and its alignment with business objectives. The process of evaluation must be ongoing and adaptive, continually refining strategies based on feedback and results. Emphasizing the dynamic nature of data culture, organizations must be agile in adapting to changing industry trends, technological advancements, and evolving organizational needs.

Leadership plays a pivotal role in the success and sustainability of a data culture. Executives and managers serve as advocates for data-driven decision-making and must continuously demonstrate their commitment to the data culture initiative. Leadership support is not a one-time endeavor; it requires sustained engagement, fostering a culture that values data and promotes its integration into all aspects of the organization.

To ensure the long-term success of the data culture, organizations must encourage ongoing commitment to data-driven practices and a data-centric mindset. This involves providing continuous learning opportunities to enhance data literacy and skills development. Organizations should also foster a culture of data sharing, collaboration, and recognition, where data-driven contributions are celebrated and employees are incentivized to embrace data-driven decision-making.

Building and sustaining a data culture is a transformative journey that requires a combination of strategic planning, leadership commitment, and a continuous learning mindset. By aligning metrics with organizational objectives, tracking data literacy and skills development, evaluating data-driven decision-making, and assessing data culture adoption, organizations can gauge the success of their data culture initiatives.

Continuous refinement of the data culture strategy based on feedback and results enables organizations to remain relevant and agile in a rapidly changing business landscape. Leadership plays a crucial role in advocating for data-driven practices and empowering data culture champions. Encouraging ongoing commitment to data-driven practices and mindset creates a thriving data culture that empowers employees, drives innovation, and positions the organization for sustained success.

As organizations embrace the data-driven future, those that prioritize and sustain a data culture will gain a competitive edge, thrive amidst uncertainty, and unlock the full potential of their data assets. Through continuous evaluation, adaptation, and leadership commitment, organizations can embrace the data culture journey and harness the power of data to make informed decisions and drive transformative outcomes.

CHAPTER 10

Case Studies and Lessons Learned

Noted Swiss psychiatrist Carl Jung said, "You are what you do, not what you say you'll do." In the organizational context, employees expect the organization and its leadership to do what they say. Culturally, "what you do" is referred to as enacted values. Espoused values are "what you say you will do" (Venema, 2021). When the two are the same, the culture thrives. When they are not, the results can be devastating.

When there's a gap between espoused and enacted values, it can lead to a disconnected organization. It means the company is not walking its talk, and that can lead to undesirable effects. For one, it causes employee dissatisfaction and loss of trust. Additionally, it contributes to the lack of company success, as the objectives and goals are in line with the espoused values, but the actions and results are in line with the enacted values. It's a big part of resistance to the change proposals by leadership (University of Minnesota, 2017; University of Minnesota, 2015).

Artifacts are the tangible and visible elements of an organization that demonstrate its values. These can include things like the physical layout of the office, company policies, and even the way employees dress and interact with one another, For example, a company that values collaboration and open communication might have an open office layout with plenty of common areas for employees to gather and work together. They might also have an executive "open door" policy, where employees are encouraged to approach their managers with ideas and concerns. On the other hand, a company that values individual achievement and competition might have a more traditional office layout with private offices and cubicles. They might also have policies in place that reward individual performance and encourage competition among employees.

G. W. Griffin and D. Holcomb, *Building a Data Culture*, https://doi.org/10.1007/978-1-4842-9966-1_10

The allocation of resources to fund "espoused" aspirations such as data accessibility can be seen as an enactment artifact in the sense that it demonstrates the organization's commitment to its stated values. By allocating resources toward data accessibility, the organization is taking concrete steps to make its espoused values a reality. This can help to bridge the gap between espoused and enacted values and can also serve as a signal to employees and other stakeholders that the organization is serious about its stated values (Buckley & McCotter, 2017).

The mismatch between espoused and enacted values can occur for many reasons within a data culture. Sometimes, a change in leadership brings aspirational desires that are not congruent with the organizational culture. Sometimes, leadership does not understand the breadth of what they are asking. Other times, actors in the environment have the power, influence, and political savviness to execute an alternate agenda.

This chapter consists of two sections. The first section will explore the use of data within industries and their match to the cultures in the UFDCM. This section will offer insight to the standard data consumption pattern within an industry and how each pattern can be facilitated by the quadrants of the model. The second section of this chapter offers use cases the authors have encountered during their careers. Some of these use cases represent success, while others demonstrate less than optimal results.

Industries and the UFDCM

Industries have regulatory, competitive, and structural needs that influence their selection of a data culture. In this section, we will explore the likely data cultures for 20 industries. Companies in these industries may choose to further democratize or restrict data flow over their industry competitors. They may select more aggressive or conservative Data Usage practices. Industries are likely to adopt a Progressive culture because they are constantly evolving and changing. Leadership may choose to break from industry norms to derive competitive advantage or cross-over into other industries.

The companies listed under each culture do not imply they adhere to a specific culture but are representative of a particular industry. Also note that companies like Apple, Inc. may fit into multiple industries and have business units that fit each of the different industries and are listed multiple times. The data cultures in these business units likely differ significantly based on the dynamics explained throughout this book.

Progressive Culture

Industries are likely to adopt a Progressive culture because they are constantly evolving and changing. Companies in these industries must be able to adapt quickly to new trends and developments in order to remain competitive. This means that they are often early adopters of new technologies and are willing to experiment with new ways of doing things. Consider the nature of startups, creative industries such as media and advertising, and constant developing industries such as technology and consulting; in each case, the level of change forces the companies to exploit data assets fully or fall behind the competition.

1. Technology Companies: These companies focus on the development and production of technology products and services. They are often at the forefront of innovation and are responsible for many of the technological advancements that have transformed our daily lives. Some of the large technology companies that might fit this category include Apple Inc., Samsung, Foxconn, Alphabet Inc., Microsoft, Huawei, Dell Technologies, Hitachi, IBM, and Sony.

2. Startups: These are newly established companies that aim to develop innovative products or services and grow rapidly. They can be found in a wide range of industries, from technology to healthcare to consumer goods.

3. Media Companies: These companies produce or distribute information, news, and entertainment content. They can range from large multinational corporations to small independent studios. Some of the large media companies that might fit this category include Apple, Disney, and Comcast.

4. Advertising Agencies: These companies create and manage advertisements and marketing campaigns for their clients. They can range from large multinational corporations to small boutique agencies. Some of the large advertising companies that might fit this category include WPP, Omnicom Group, Publicis Groupe, Interpublic Group of Companies, and Dentsu.

5. Consulting Firms: These firms provide expertise and specialized labor to their clients through the use of consultants. They can range from large multinational corporations to small boutique firms. Some of the large consulting companies that might fit this category include McKinsey & Company, Bain & Company, Boston Consulting Group, Deloitte Consulting, and PwC Consulting.

Traditional Culture

Industries are likely to adopt a Traditional culture when they have been around for a while or follow established ways of doing things. Industries like retail, consumer goods, manufacturing, transportation, and hospitality have large hierarchies and specialized processes that can make it difficult for them to adapt quickly to new trends or technologies but allow them to easily share measurement throughout the organization.

Although these industries touch the broad citizenry, service disruption presents less immediate concern than other industries:

1. Retail Companies: These companies sell goods directly to consumers through various channels such as brick-and-mortar stores, online stores, or catalogs. They can range from large multinational corporations to small independent shops. Some examples of the large retail companies that might fit this category include Walmart Inc., Amazon.com Inc., Costco Wholesale Corporation, The Kroger Co., Walgreens Boots Alliance Inc., The Home Depot Inc., CVS Health Corporation, Target Corporation, Lowe's Companies Inc., and Albertsons Companies Inc.

2. Consumer Goods Companies: These companies manufacture and sell products for personal or household use. These products can range from food and beverages to clothing and electronics. Some of the large consumer goods companies that might fit this category include Procter & Gamble Co., Unilever PLC/NV, Nestle S.A., PepsiCo Inc., and The Coca-Cola Co.

3. Manufacturing Companies: These companies produce goods from raw materials or components using machines or labor-intensive methods such as assembly lines or handcrafting techniques. Some of the large manufacturing companies that might fit this category include Apple, Toyota Group, Volkswagen Group, Samsung Electronics, and Foxconn.

4. Transportation Companies: These companies provide transportation services for people or goods. They can range from large multinational corporations to small independent operators. Some of the large transportation companies that might fit this category include United Parcel Service Inc., FedEx Corporation, Delta Air Lines Inc., American Airlines Group Inc., and Southwest Airlines Co.

5. Hospitality Companies: These companies provide lodging, food, and other services to travelers. They can range from large multinational corporations to small independent operators. Some of the large hospitality companies that might fit this category include Marriott International Inc., Hilton Worldwide Holdings Inc., Wyndham Hotels & Resorts Inc., InterContinental Hotels Group PLC, and Accor S.A.

Preservationist Culture

An industry is likely to adopt a Preservationist culture because it is heavily regulated and has strict standards for safety, efficacy, quality, and sometimes environmental protection. Healthcare, financial, pharmaceutical, and energy companies must adhere to these regulations and standards in order to operate, which can make it difficult for them to adopt new technologies or approaches quickly. These five industries have unique profiles:

1. Healthcare Companies: These companies provide medical goods and services to individuals and organizations. They can range from large multinational corporations to small independent clinics. Some of the large healthcare companies include UnitedHealth Group Incorporated, CVS Health Corporation, AmerisourceBergen Corporation, Cardinal Health Inc., and McKesson Corporation.

2. Financial Services Companies: These companies provide financial products and services to individuals and organizations. They can range from large multinational banks to small independent financial advisors. Some of the large financial services companies that might fit this category include JPMorgan Chase & Co., Bank of America Corporation, Wells Fargo & Company, Citigroup Inc., and Goldman Sachs Group Inc.

3. Insurance Companies: These companies provide insurance products to protect individuals and organizations from financial losses. They can range from large multinational corporations to small independent agencies. Some of the large insurance companies that might fit this category include UnitedHealth Group Incorporated, Anthem Inc., Humana Inc., Centene Corporation, and Cigna Corporation.

4. Pharmaceutical Companies: These are independent pharmaceutical, biotechnology, and medical companies that have generated a revenue of at least US$10 billion. Some of the large pharmaceutical companies that might fit this category include Johnson & Johnson, Roche Holding AG, Pfizer Inc., Novartis AG, and Merck & Co. Inc.

5. Energy Companies: These companies produce or distribute energy in various forms such as electricity, natural gas, or oil. They can range from large multinational corporations to small independent operators. Some of the large energy companies that might fit this category include ExxonMobil Corporation, Chevron Corporation, ConocoPhillips Company, Phillips 66 Company, and Valero Energy Corporation.

Protectionist Culture

Industries select the Protectionist culture due to the sensitive nature of the data being handled and the interest of citizenry inherent in their products. This means that they must be cautious in their approach to new technologies or approaches and must carefully consider the potential risks and benefits before making any changes.

For example:

- Government agencies are likely to adopt a Protectionist culture because they are responsible for protecting the interests of their citizens and maintaining law and order.

- Defense contractors are responsible for providing critical products and services to the military or intelligence departments.

- Public utilities are responsible for providing essential services to their citizens.

- Mining companies are responsible for extracting valuable resources from the earth.

- Agricultural companies are responsible for providing food and other essential products to their citizens.

These industries have unique profiles marked by their touch of the broad citizenry:

1. Government Agencies: These are permanent or semi-permanent organizations within the government that are responsible for the oversight and administration of specific functions. They can range from large federal agencies such as the Department of Defense or the Environmental Protection Agency to smaller local agencies.

2. Defense Contractors: These are business organizations or individuals that provide products or services to a military or intelligence department of a government. Products typically include military or civilian aircraft, ships, vehicles, weaponry, and electronic systems, while services can include logistics, technical support and training, communications support, and engineering support in cooperation with the government. Some of the large defense contracts that might fit this category include Lockheed Martin, Boeing, and Raytheon Technologies in the United States. In other countries, examples include BAE Systems in the United Kingdom, Thales Group in France, and Leonardo S.p.A. in Italy.

3. Public Utilities: These are organizations that maintain the
    infrastructure for a public service (often also providing a
    service using that infrastructure). Public utilities are subject
    to forms of public control and regulation ranging from local
    community-based groups to statewide government monopolies.
    Public utilities are meant to supply goods/services that are
    considered essential; water, gas, electricity, telephone, and
    other communication systems represent much of the public
    utility market.

4. Mining Companies: These are businesses that extract minerals
    from the earth. They can range from large multinational
    corporations to small independent operators. Some of the large
    mining companies that might fit this category include Rio Tinto,
    BHP Billiton, and Anglo American.

5. Agricultural Companies: These companies focus on growing
    plants and raising animals for commercial uses as food and fiber.
    They can range from large multinational corporations to small
    independent farms. Some of the large agricultural companies that
    might fit this category include Monsanto (now owned by Bayer),
    Archer Daniels Midland, and Cargill.

Case Studies and Experiences

This section will offer examples from the authors' experience. We provide this section
to our readers to help them use the material contained throughout previous chapters of
this book. Each section will give an overview of the situation and draw upon the UFDCM
to add context to the results received. As we take matters of privacy and confidentiality
very seriously, we have intentionally left out names that may identify specific people
and the organization. The case studies provided are real with real people and real
situations. Although the ideal of building a data culture offers many substantial benefits,
the practical realities are more often full of challenges. The struggle to leverage data as a
strategic asset is real.

Case Study #1

Background: Organization W (Org W) had a mature balanced scorecard system, which was reported three times a year to the board and the public. However, the Local Education Agency (LEA) lacked a data warehouse, an enterprise data model, and enterprise data standards. This absence of data governance contributed to serious data quality issues within the organization. Furthermore, the LEA's leader, a demanding individual who led with an iron fist, perpetuated an oppressive data culture. Data was used as a battering ram, leveraging it to control and punish individuals rather than using it as an instrument for continuous improvement.

Crisis and Investigation: Amidst this challenging data landscape, the Local Education Agency was accused of cheating on state standardized tests by the State's Department of Education and the Governor's office. The accusations led to a thorough investigation, revealing instances of test result manipulation and falsification within the LEA. Several individuals were prosecuted, further damaging the LEA's reputation and exacerbating the oppressive data culture.

Transition to a Data-Driven Environment: Recognizing the need for change, Org W sought external expertise to establish a data governance program and shift its data culture toward a more positive and empowered approach. The establishment of a data governance program was proposed and accepted by the Senior Cabinet consisting of all C-level decision-makers within the LEA.

1. Data Inventory: As part of the recommended data governance program, the first step was to conduct a comprehensive data inventory. This involved identifying and documenting all data sources, data flows, and data stakeholders within the LEA. The data inventory aimed to establish a baseline understanding of the LEA's data landscape and serve as a foundation for future data governance initiatives.

2. Leadership Alignment: To initiate cultural change, it was essential to align the LEA's leadership team around the vision of a data-driven environment. This involved extensive communication and education on the benefits of a collaborative and transparent data culture, focusing on the value of data as a catalyst for continuous improvement and student success.

3. Establishing Data Governance Framework: Org W implemented a robust data governance framework to govern data management, data quality, data privacy, and data security. This framework included the development of a data warehouse, an enterprise data model, and enterprise data standards. Clear policies and procedures were put in place to ensure the responsible and ethical use of data throughout the LEA.

4. Training and Capacity Building: To empower stakeholders, the LEA provided comprehensive training and capacity-building initiatives on data literacy, data analysis, and data-driven decision-making. This included tailored programs for educators, administrators, and support staff, enabling them to leverage data effectively and contribute to a data-driven culture.

5. Cultivating Collaboration and Continuous Improvement: The LEA fostered a collaborative environment, encouraging cross-functional teams to analyze data, identify improvement opportunities, and implement evidence-based interventions. Regular data-driven discussions were held to share insights, celebrate successes, and drive continuous improvement across the organization.

Results and Impact: Over time, the transformation to a data-driven environment yielded significant improvements. The LEA experienced enhanced data quality, increased transparency, and improved decision-making based on reliable and accurate information. Collaboration and trust among stakeholders were strengthened, leading to a more cohesive and empowered workforce.

Conclusion: Org W's journey toward a data-driven environment illustrates the importance of addressing data governance, quality, and the misuse of data as a tool for control. By establishing a robust data governance program, fostering collaboration, and promoting a positive data culture, the LEA was able to leverage data as a valuable asset for continuous improvement and student success. This case study serves as a valuable example for organizations seeking to transform their data culture and unlock the full potential of data-driven decision-making.

The organization's data culture can be classified as Protectionist according to the Usage and Flow Data Culture Model. A Protectionist data culture emphasizes data security, privacy, and compliance, which is not surprising given the protective nature of this government organization. This culture is concerned with safeguarding data assets and mitigating risks associated with data misuse or breaches. Org W demonstrates Protectionist characteristics. As they were accused of cheating on state standardized tests, there was a clear need for data security and integrity, as well as data literacy along with the development of mature processes that were regularly monitored. Org W's data culture maturity scorecard is shown in Table 10-1.

Table 10-1. *Organization W's Data Culture Maturity Scorecard*

Data Culture Scorecard – Organization W		
Component	**Status**	**Comments**
Data Strategy	No Formal Data Strategy	Integration layer – basic level plan Data Movement – basic level plan DW infrastructure – No enterprise data warehouse
Data Governance Office	No Central Structure	No central place to go
Data Governance Program	Proposed and accepted Initial StageP	Librarian – Early (Metrics only) Data Quality – Ad hoc Process – None Compliance – Ad hoc / regulatory Communication – Ad hoc Training – no coordination
Operational and Data Processes	Operational – initial stage Data – Ad Hoc	Operational process improvement in early stage No enterprise processes documented in central repository Data Flows did not exist
Data Literacy	None	No formal training program
Data Analytics	No single reporting group. Different depts for different reporting reqs	Results differed between groups
Data Privacy and Security	No CISO	Focused on explicit State and Federal regulatory requirements

Case Study #2

Background: This case study explores the data culture of Organization X (Org X), a publicly traded company in the freight and logistics industry, during its ongoing five-year digital transformation. We analyzed the organization's data management practices and assessed its alignment with the Usage and Flow Data Culture Model. Additionally, we examined the challenges faced by the enterprise in establishing a modern data platform and fostering a data-driven environment.

1. Introduction: Org X operated in the freight and logistics industry and had embarked on a significant enterprise digital transformation initiative. All data management and reporting activities for the enterprise resided within a central group located in the Information Technology department. The primary focus of the transformation involved modernizing the company's application infrastructure and supporting the development of custom applications. This case study examined the data culture of Org X, focusing on its data management practices, technology landscape, and challenges encountered during the digital transformation.

2. Background: Org X's digital transformation aimed to streamline operations, improve efficiency, and enable innovative solutions through technology adoption with many new custom applications. All reporting and business intelligence involved a small data solutions department, reporting to a director-level role. The initial task was the creation of a data model for an operational data store (ODS) that aligned with an enterprise service bus architecture. The organization utilized API development and management for data transformation.

3. Data Management Landscape: Org X's data management landscape was characterized by several gaps and limitations. The absence of metadata management, an enterprise data model, enterprise data dictionary, and organizational chart posed challenges to comprehensive data governance. Furthermore, there was a lack of enterprise business process management and improvement initiatives. Despite these constraints, proactive measures were taken to address some of these gaps by setting up a data catalog, documenting over 500 business processes, and creating a glossary with over 900 terms.

4. Data Modeling Efforts: As part of the digital transformation, over 100 physical and logical data models were also added to the data catalog. This included evaluating and incorporating an enterprise data model, which was assessed against the organization's

use cases and encompassed both new builds and legacy data models. Although there were endeavors to establish robust data management practices, senior management within the IT department did not fully support the rollout of the data catalog, impeding its adoption on the enterprise level.

5. Funding Constraints and Data Warehouse Modernization: Despite the comprehensive digital transformation efforts, no comprehensive funding was allocated to modernize the organization's 20-year-old data warehouse solution. This financial constraint posed challenges to achieving a truly data-driven environment, hindering the organization's ability to fully leverage data as a strategic asset.

6. Classification Based on the Usage and Flow Data Culture Model: Considering the organization's data management practices and its approach to the digital transformation journey, Org X's data culture can be classified as Traditional within the Usage and Flow Data Culture Model. A Traditional data culture is characterized by a reliance on established practices and hierarchical decision-making processes. This culture lacked a proactive approach to data management and innovation. While Org X's leader exhibits a Traditional leadership style by exerting control, the organization's efforts to build a data governance program and embrace a data-driven environment suggest a departure from a Traditional data culture. While efforts were made to introduce data cataloging, document business processes, and enhance data modeling, the limited support for the data catalog rollout, absence of metadata management, and outdated data warehouse solution clearly is best described as a Traditional data culture.

7. Conclusion: This case study sheds light on Org X's data culture during its ongoing digital transformation. Despite the organization's Progressive aspirations through digitalization, several limitations and challenges hindered the establishment of a fully Progressive data culture. The findings emphasize the importance of comprehensive data management practices,

funding for modernization efforts, and senior management support in fostering a data-driven environment. As shown in Table 10-2, Org X's data culture still has considerable work to completely establish a mature data culture.

Table 10-2. *Organization X's Data Culture Maturity Scorecard*

Data Culture Scorecard – Organization X		
Component	**Status**	**Comments**
Data Strategy	No Formal Data Strategy	Integration layer – advance level plan - legacy Data Movement – Intermediate level plan - legacy DW infrastructure – Legacy Data Platform Modernization Effort Early
Data Governance Office	No Central Structure	Data Governance Ad hoc Reporting / BI / Analytics in IT Dept.
Data Governance Program	No Formal Program	Librarian – Intermediate: Data Catalog, Business Glossary Data Quality – Advanced for Legacy DW Process – No formalized and published processes Compliance – Ad hoc / regulatory Communication – Ad hoc Training – no coordination
Operational and Data Processes	Operational – initial stage Data – Ad Hoc	Operational process were being documented and added to Data Catalog Repository; Data – Flow were legacy around DW
Data Literacy	None	No formal data literacy training program
Data Analytics	Single BI/Reporting Group in IT	Legacy DW Reporting/BI – Advanced Little Predictive Analytics No Data Science Capabilities
Data Privacy and Security	VP Level CISO in IT	Focused on IT processes and explicit regulatory requirements

Case Study #3

Background: Organization Y (Org Y) was a publicly traded company in a regulated industry. Org Y had to adhere to governmental regulations and reporting regarding funding, industry regulatory bodies for certifications, and the military for product agreements.

Organizational Life Cycle and Organizational Culture. Org Y was in the mid-late growth phase. They still had employees from the early startup phase, and the leader moving them from the startup through the early growth phase of the organizational life cycle had recently retired.

Org Y had aggressive growth and expansion plans both organically and inorganically. After a successful acquisition and turnaround of Org Y1, a significantly smaller company than Org Y, the organization embarked on additional acquisitions and integrations to satisfy the inorganic growth goals. The new acquisitions consisted of an organization about the same size as Org Y (Org Y2) and a smaller organization (Org Y3). Org Y2 was older and a little later in the organizational life cycle than the acquirer. Org Y3 was also later in the life cycle and had a completely different origin story. Despite these differences, the organizations had similar cultures. Both were hierarchical with a family feel. They tended toward a hybrid hierarchy/clan organizational culture.

The organization's leadership also embarked on brand, product, and marketing expansion to spur growth organically. These branding efforts were made to increase appeal in new markets. The product expansion plans were directed at the current and expanded markets. The marketing efforts offered new ideas for portals and marketing lyrics. This marketing would support the branding efforts as well.

The leaders of Org Y were positioning the organization for further growth and scale by implementing process improvements, data exploitation, and differentiation. The leadership decided to organize using a central organizational structure with business units and some shared services to optimize resources. This approach allows the organization to maximize services such as technology, finance, marketing buys, and data services. To do this, the organization hired new technology and data leadership to drive the efforts.

Data Culture: The data culture in place before the acquisitions supported the organizational culture with informal processes and data handling. Manual processes, spreadsheets, and other ad hoc methods were used during the startup/growth phase. Moving the organization toward a scalable and formal process requires strong executive commitment and vision. As seen by the initial review (see Table 10-3), the components of a mature data culture had some solid elements and others needing development. Leadership took steps to move toward the targeted data culture, which would be characterized by moderate data flow to match regulatory requirements with higher levels of analytics to encourage differentiation. This selection would put the desired state on the right side of the Preservationist culture or the left side of the Progressive culture.

Key Learning: Integrating the organizational cultures of multiple business units must be a priority. Restrictions based on regulators must be respected, but the cultures should be somewhat normalized due to the corporate model selected. As the shared services are implemented, the cultural dynamics and expectations were not standardized, and the organization at large struggled with inconsistent decision-making.

From a data culture perspective, the organization desired to move to more advanced Data Usage. The data flow from the business units to the newly formed shared services organization requires expanded but still limited data flow. The movement from informal to formal scalable processes requires executive leadership to support good governance and model the behaviors. This support must be visible and sustained.

The Org Y use case demonstrates the power of the components of a mature data culture in cultural transformation. First, the components of a mature data culture can be implemented regardless of the culture selected. The components act as a toolkit for establishing and sustaining your desired culture. So for Org Y, each element should be enhanced for all business units and supported during the organizational culture shift. These components can increase the speed of modifying decision-making and create the consistency needed.

The current component evaluation (see Table 10-3) offers a road map to infrastructural needs. The data strategy was an accepted practice and should be leveraged to acquire funding for the data infrastructure. The organization had made hires to build a data governance office, consolidate the reporting functions in Org Y, and begin to promote the data office. These changes took place during the acquisitions of Org Y2 and Org Y3. These acquisitions required additional infrastructure to support improvements in business unit operations. Unfortunately, the organization did not include process management in the funding or planning. This lack of process management slowed progress of integration, new product development, product launches, and data exploitation.

Data literacy was a bright spot in the organization. By setting up knowledge-sharing events with various organizations, Org Y increased data literacy around the meaning of data elements and the proper handling of data for regulatory purposes. Training in multiple departments was initiated for data literacy and quality initiatives. Individuals began to volunteer for data stewardship roles and training roles within and across departments.

What happened? Org Y introduced data governance activities to improve data quality in multiple business units, including data cleanup in preparation for master data management. Other initiatives included building data warehouses and reporting infrastructures, increasing self-service, supporting personalization in user portals, and engaging outside consultants to create advanced behavioral models. The organization could have benefited from process mapping and management, better data flow analysis, enhanced data warehousing capabilities, standardizations of reporting needs across business units, and increased usage of advanced analytics techniques to understand buyer's behaviors.

Org Y struggled with merger and acquisition integration, translating into difficulty in building a cohesive organizational and data culture. The organizational structure was modified several times with little notable traction. Eventually, leadership centralized decision-making to reduce variance and ensure data was properly interpreted. Unfortunately, gatekeepers to leadership required reporting and data manipulation, reducing transparency between the operational and key performance data. This manipulation was intended to focus results and, in some cases, did. The difficulty came when initiatives claimed levels of effect that could not be proven. The problem with its organizational culture transformation directly impacted the data culture transitions.

Org Y is a great case of an organizational culture in flux due to its place in the organizational life cycle. This flux became more difficult to address due to a lack of true integration of previous acquisitions and the continued activity in the acquisition space. These situations affected the establishment of a data culture and the use of data. Leaders should always recognize the dynamics of their place in the organizational life cycle, evaluate integrations of cultures, strive to determine the corporate culture they wish to have, and enhance the components of a mature data culture to move to and sustain the desired data culture.

Table 10-3. *Organization Y's Data Culture Maturity Scorecard*

Data Culture Scorecard – Org Y		
Component	Status	Comments
Data Strategy	IT had a data strategy	Integration layer – advance level plan Data Movement – basic level plan DW infrastructure – moderate (Business Unit based)
Data Governance Office	No Central Structure	Data Governance Ad hoc COE Reporting / BI / Analytics – distributed – no central point
Data Governance Program	Ad hoc	Librarian – Early (Metrics only) Data Quality – Ad hoc Process – In Project Documentation Compliance – Ad hoc / regulatory Communication – Ad hoc Training – no coordination
Operational and Data Processes	Operational – Project Docs Data – Ad Hoc	Operational process are documented as part of each project; no central process repository; Data – Process data flows do not exist; DW: moderate
Data Literacy	None	Only to meet regulatory requirements
Data Analytics	Multiple reporting groups	Results differed between groups and timing
Data Privacy and Security	Recently hired CISO	Focused on IT processes and explicit regulatory requirements

Case Study #4

Background: Organization Z (Org Z) is a technology product startup in the cyber industry.

Organizational Life Cycle and Organizational Culture: As a startup, Org Z must have the information to allow founders to make decisions on burndown rates, pipelines, etc. Generally, the organizational culture of a startup is built around the founders and becomes a family-type environment or "clan" culture. The clan organization is high in collaboration and inclusion. With few members, the culture and decision-making rights center around the founders, funding sources, and boards. In Org Z, the leadership was given broad latitude to get things done by the founders and board.

Startups generally run on a minimalistic approach with a focus on the product. The term MVP, or minimum viable product, epitomizes the startup culture. The product can evolve based on the reception of the marketplace to the MVP. The product features can change based on this reception and feedback. Similarly, the operational aspects must be put into place as little exists. A startup shows the strength of the seven-component model.

Although we listed data strategy first in the components list, Org Z started with process management. Like all startups, Org Z began with very few defined processes (see Table 10-4). These process definitions took two tracks. The first was the product track, including feature approval, software development life cycle, system capacity, and usage. The second was the operational track, which includes the sales process (acquisition and renewal), customer onboarding, provisioning, marketing, security, etc. This development included the leaders from each domain and was overseen by the chief operating officer.

Table 10-4. *Organization Z's Data Culture Maturity Scorecard*

Data Culture Scorecard – Org Z		
Component	Status	Comments
Data Strategy	Ideation	Startup with little data strategy direction; no data strategy existed
Data Governance Office	None	Formal data governance was not considered; product development was the focus; product development as early stage and somewhat ad hoc with few governance, especially data governance, processes
Data Governance Program	None	Librarian – Ad hoc Data Quality – early stages Process – Little established (Grew to mature) Compliance – Regulatory / advanced Communication – Ad hoc Training – moderate – regulatory
Operational and Data Processes	None	Operational Processes – founders led Data Process – new / not established
Data Literacy	None	Data Handling was coupled with cyber security training – High Data Usage – minimal
Data Analytics	Part of product – Advanced Business – Light	Product was an analytics process and had advanced functionality Business processes were light in analytics
Data Privacy and Security	Advanced – because of industry	Because the industry deal t with cyber security and risk analysis, Data privacy practices were advanced (lots of training)

With the processes in place, the organization could ensure data was collected at the appropriate points. The organization began to evaluate marketing and contact programs, sales pipelines, and product development. In the case of product development, sign-off processes, release schedules, and testing processes were documented and implemented. The insights from this documentation effort accelerated feature implementation dramatically. Similar enhancements in the customer onboarding process were realized as measurements were implemented.

Data Culture: The nature of the startup placed a high value on using analytics while maintaining a high level of security. The flow of client data was limited to ensure none was inadvertently exposed. The use of analytics was to continuously evaluate and understand the clients environment for cyber resiliency. The analytics used would be considered advanced. On the internal or operational data, the data flowed based on need. Within the mid- to senior levels, data flowed freely regarding the operations, marketing, and sales functions. Little data was shared outside of the organization. This combination of data flow being limited and Data Usage being moderate to advanced would put the data culture in the rightmost section of the Preservationist quadrant or the leftmost section of the Progressive quadrant.

Learning and Complexities: Org Z shows the advantages of establishing processes in a startup. The processes can be optimized to collect data. This optimization allows the organization to maximize the data for operational efficiency. Additionally, the creation of processes and the communication of those processes and their interdependencies increase collaboration between functions in the organization. A key to process management in a startup is engaging the founders. These visionaries are looking to advance their product, and having these processes documented and measured frees them to act. This engagement can be difficult because it requires some rigidity, so getting their buy-in and approvals is important.

Org Z was unique in its recognition of data and its potential. The founders came from data-centric backgrounds and understood the need to exploit the data asset fully. With this understanding also came a respect for data privacy and security. Rarely do you have this combination in the senior leadership. This combination led to the creation of the startup and, ultimately, its culture.

From an organizational design perspective, the startup is rather simple. From a data perspective, there were many complexities with client data and its configuration, external entities and their data nomenclature, and the various frameworks associated with industry or governing bodies. The data landscape was broad, which made the need for Data Flow documentation exceptionally important. Not all data had the same velocity, and some changes to external data obtained through APIs (application programming interfaces) were ad hoc with little communication. These challenges required significant management and feedback loops in the processes to ensure data quality. The simplicity of a startup's organizational design and the autonomy in this particular organization allow these processes and feedback loops to be put into place quickly.

Summary

In the culminating chapter of this book, we cover the intricacies of building a data culture through the lens of real-world experiences and case studies. As Swiss psychiatrist Carl Jung aptly noted, actions speak louder than words, and the alignment between an organization's espoused and enacted values is a crucial determinant of its success. This chapter underscores the significance of these values within the realm of data culture and how they can impact an organization's journey toward a mature data-driven mindset.

The chapter commences by exploring the critical interplay between espoused and enacted values, emphasizing the potential consequences of a disconnection between the two. Such a divergence can lead to a host of issues, including employee dissatisfaction, eroded trust, and diminished company performance. To illuminate this concept, we introduce the notion of artifacts as tangible manifestations of organizational values, encompassing everything from office layout to dress codes. A compelling illustration is provided wherein organizations valuing collaboration cultivate open office spaces and promote an "open door" policy to foster employee interaction.

A pivotal aspect of closing the gap between espoused and enacted values lies in the allocation of resources that align with the organization's professed values. This section explores how this resource allocation, particularly in relation to data accessibility, serves as an enactment artifact. By investing in data accessibility, organizations transform aspirations into tangible reality, signalling their commitment to stated values. This dynamic plays a pivotal role in harmonizing organizational culture, minimizing the dissonance between what is said and what is done.

The chapter proceeds with a comprehensive exploration of the correlation between industries and data culture. By analyzing the unique regulatory, competitive, and structural dynamics within industries, we shed light on the data cultures that emerge. Industries such as technology, media, and advertising, often characterized by constant evolution and change, tend to adopt a Progressive data culture, staying at the forefront of innovation. Meanwhile, well-established industries like retail, consumer goods, and manufacturing may lean toward a Traditional data culture, valuing stability and established processes. Highly regulated sectors such as healthcare and finance often adopt a Preservationist data culture to adhere to strict standards. Finally, organizations with large amounts of sensitive data or that cater to the citizenry such as government and public utilities often select the Protectionist culture to ensure privacy and security.

The chapter culminates with a series of enlightening case studies drawn from the authors' firsthand experiences. These real-world examples offer readers valuable insights into the complexities of building a data culture, highlighting successes as well as challenges. While the concept of a data-driven environment brings substantial benefits, the authors acknowledge the practical hurdles organizations face in their pursuit of a mature data culture. In this final exploration, the chapter serves as a bridge between theory and practice, enriching readers with practical wisdom and tangible instances to grasp the nuances of implementing a data culture. The experiences shared underscore the fact that the journey toward a data-driven environment is marked by real-world challenges and triumphs, offering a balanced and comprehensive perspective that equips readers to navigate their own unique paths to success.

Bibliography

Abdul-Jabbar, S. S. and Farhan, A. K. (2022). Data Analytics and Techniques: A Review. ARO-The Scientific Journal of Koya University, X(2), ARO.10975. 11 pages.

AccelData (n.d.). What is Data Culture? Retrieved July 26, 2023, from www.acceldata.io/article/what-is-data-culture.

Aiello, L. (August 1, 2019). Culture + technology + data = employee engagement. HRD Connect. Retrieved July 26, 2023, from www.hrdconnect.com/2019/08/01/culture-technology-data-employee-engagement/.

American Heritage Dictionary. (2023). Culture. American Heritage Dictionary Online. www.ahdictionary.com/word/search.html?q=culture.

Anderson, J. (2020). Data Teams: A Unified Management Model for Successful Data-Focused Teams. Apress.

Azhar, A. (September 8, 2022). Data literacy: What is it and why is it essential for success? Retrieved June 15, 2023, from www.techrepublic.com/article/what-is-data-literacy/.

Borner, I. (January 25, 2023). Creating a Culture of Privacy: The Importance of Developing a Privacy Program For Your Business. Forbes Business Council COUNCIL POST. Retrieved August 7, 2023, from www.forbes.com/sites/forbesbusinesscouncil/2023/01/25/creating-a-culture-of-privacy-the-importance-of-developing-a-privacy-program-for-your....

Brower, R. L., Mokher, C. G., Jones, T. B., Cox, B. E., and Hu, S. (Jan–Mar 2020). From Democratic to "Need to Know": Linking distributed leadership to data culture in the Florida College System. AERA Open (6)1, https://doi.org/10.1177/2332858419899065.

BrainyQuote (n.d.). Piyush Goyal Quotes. Retrieved from www.brainyquote.com/authors/piyush-goyal-quotes.

Bulkley, K. E. and McCotter, S. S. (2017). Learning to Lead with Data: From Espoused Theory to Theory-in-Use. Leadership and Policy in Schools, 17(4), pp. 591–617. https://doi.org/10.1080/15700763.2017.1326144.

Cameron, E. and Green, M. (2015). Making sense of change management: A complete guide to the models, tools, and techniques of organizational change. London, UK: Kogan Page.

Cameron, K. S. and Quinn, R. E. (2011). Diagnosing and changing organizational culture: based on the competing values framework. Jossey-Bass.

Cinnamon, J. (2022). On data cultures and the prehistories of smart urbanism in "Africa's Digital City." Urban Geography. Ahead of print, 1–22. https://doi.org/10.108 0/02723638.2022.2049096.

Cortellazzo, L., Bruni, E., and Zampieri, R. (2019). The Role of Leadership in a Digitalized World: A Review. Frontiers in Psychology, 10, 1938. doi: 10.3389/ fpsyg.2019.01938.

Correlation One (May 4, 2023). How Digital Leaders Can Foster a Culture of Continuous Learning. Retrieved July 26, 2023, from www.correlation-one.com/blog/ continuous-learning.

Covey, S. R. (2004). The 7 habits of highly effective people: Restoring the character ethic [Rev. ed.]. Free Press.

DAMA International (2017). DAMA-DMBOK: Data Management Body of Knowledge (2nd ed.). Technics Publications.

Data and Analytics Group (July 3, 2023). The Key to Data Leadership: Enabling Culture Change and Technology Adoption in the C-Suite. Retrieved July 26, 2023, from www.linkedin.com/pulse/key-data-leadership-enabling-culture-change/.

Dictionary.com (2023). Culture. Dictionary.com. www.dictionary.com/browse/ culture.

Domo (n.d.). The role leaders can play in building a data-driven culture. Retrieved July 26, 2023, from www.domo.com/learn/article/leaders-role-in-building-data- driven-culture.

Eads, A. (February 3, 2023). What Is Organizational Communication? (Types and Importance). Indeed.com. Retrieved from www.indeed.com/career-advice/career- development/organized-communication.

Edwards, J. (June 16, 2023). 7 enterprise data strategy trends. CIO. Retrieved from www.cio.com/article/412908/7-enterprise-data-strategy-trends.html.

Edwards, J. (2018). Data Silos: Now and Forever? InformationWeek. Retrieved from www.informationweek.com/it-strategy/data-silos-now-and-forever-

Eiloart, J. (January 16, 2020). Data champions are the backbone of data culture. DataIQ. Retrieved July 26, 2023, from www.dataiq.global/articles/articles/data- champions-are-the-backbone-of-data-culture.

Enzensberger, H. M. (n.d). Quotes on Culture. www.commisceo-global.com/blog/ what-is-your-favourite-quote-defining-culture.

Fisher, T. (2009). The Data Asset: How Smart Companies Goven Their Data for Business. John Wiley and Sons, Inc.

Flores, W. (August 26, 2016). Boost Your Data-Driven Culture with Data Advocacy. TDWI Website. Accessed on July 26, 2023, from `https://tdwi.org/articles/2016/08/26/boost-your-data-driven-culture-with-data-advocacy.aspx`.

Frisk, J. E. and Bannister, F. (2017). Improving the use of analytics and big data by changing the decision-making culture. Management Decision, 55(10), 2074–2088. DOI 10.1108/MD-07-2016-0460.

Gardner, R. (n.d.). 12 Types of Organizational Culture and HR's Role in Shaping It. AIHR: Academy of Innovative HR. `www.aihr.com/blog/types-of-organizational-culture/`.

Gartner (n.d.). Data Literacy as a Factor for Data-Driven Cultural Change in an Organization. Retrieved July 26, 2023, from `https://hyperight.com/data-literacy-as-a-factor-for-data-driven-cultural-change-in-an-organization/`.

Ghosh, P. (June 22, 2021). The Future of Data Literacy. Retrieved June 15, 2023, from `www.dataversity.net/the-future-of-data-literacy/`.

Gummer, E. S. and Mandinach, E. B. (2015). Building a Conceptual Framework for Data Literacy. Teachers College Record, 117(040305).

Holcomb, D. (May 6, 2010). Simplicity and Transparency: How to do effective data warehousing and business intelligence. Business Intelligence Symposium. University of Northern Kentucky & Lucrum. `www.slideshare.net/lucrum/bringing-data-to`.

Holcomb, D. D. (2009). An extension of leader-member exchange (LMX) beyond the direct manager dyad and their correlations to the member's organizational commitment. (PhD Dissertation). United States. California. TUI University.

Holcomb, D. D. (2022). A phenomenological study exploring remote workers' use of technology to replace the in-office experience. (PhD Dissertation). United States. Kentucky. University of the Cumberlands.

Hollister, R., Tecosky, K., Watkins, M., and Wolpert, C. (August 10, 2021). Why Every Executive Should Be Focusing on Culture Change Now. MIT Sloan Management Review. Retrieved July 26, 2023, from `https://sloanreview.mit.edu/article/why-every-executive-should-be-focusing-on-culture-change-now/`.

IBM (2023). What Is Data Governance? Retrieved from `www.ibm.com/topics/data-governance`.

Jackson, Y. (n.d.) *Encyclopedia of Multicultural Psychology*, p. 203; as referenced by Wikipedia (2023).

Jung, C. G. (n.d.). Quotes. Goodreads.com. Retrieved from www.goodreads.com/quotes/3240.

Kahn, S. D. and Koralova, A. (2022). A journey toward an open data culture through the transformation of shared data into a data resource. Data & Policy, 4, e29. doi:10.1017/dap.2022.22.

Kaplan, R. S. and Norton, D. P. (1992). The Balanced Scorecard – Measures that Drive Performance. Harvard Business Review, 70(1), 71–79.

Kenton, W. (September 4, 2020). Silo Mentality: Definition in Business, Causes, and Solutions. Retrieved from www.investopedia.com/terms/s/silo-mentality.asp#:~:text=In%20business%2C%20organizational%20silos%20refer,shared%20because%20of%20system%20limitations.

Kesari, G. (May 31, 2022). Does Your Data & Analytics Strategy Have These 10 Crucial Elements? Forbes. Retrieved June 16, 2023, from www.forbes.com/sites/ganeskesari/2022/05/31/does-your-data--analytics-strategy-have-these-10-crucial-elements/?sh=45291f641b24.

Kotter, J. P. (1996). Leading change. Boston, MA: Harvard Business School Press.

Lasater, K., Alabiladi, W. S., Davis, W. S., and Bengtson, E. (October 2020). The data culture continuum: An examination of school data cultures. Educational Administration Quarterly, 56(4), 533–569. https://doi.org/10.1177/0013161X19873034.

Linke, R. (Sep 14, 2017). Design thinking, explained. *MIT Sloan School of Management.* https://mitsloan.mit.edu/ideas-made-to-matter/design-thinking-explained.

Machlup, F. (1962). The production and distribution of knowledge in the United States. Princeton, NJ: Princeton University Press.

Marr, B. (2017). Data Strategy: How to profit from a world of big data, analytics, and the internet of things. Great Britain and the United States: Kogan Page Limited.

MasterClass (August 30, 2022). Data Strategy Definition: 7 Key Elements of Data Strategy. Retrieved June 16, 2023, from www.masterclass.com/articles/data-strategy.

Mayer-Schönberger, V. and Cukier, K. (2013). Big data: A revolution that will transform how we live, work, and think. Boston, MA: Houghton Mifflin Harcourt.

McAfee, A., Brynjolfsson, E., Davenport, T. H., Patil, D. J., and Barton, D. (2012). Big data: The management revolution. Harvard Business Review, 90(10), 60–68.

McKeen, J. D. and Smith, H. A. IT Strategy: Issues and Practices, Third Edition. Pearson, 2015, ISBN-13 978-0-13-354424-4.

McCoy, J. (n.d.). Organizational life cycle: Definition, models, and stages. AIHR: Academy of Innovative HR. www.aihr.com/blog/organizational-life-cycle/.

MIT Sloan (January 23, 2023). Data literacy for leaders. Retrieved June 15, 2023, from `https://mitsloan.mit.edu/ideas-made-to-matter/data-literacy-leaders`.

Nash, J. C. (July 17, 2018). Data literacy and data science literacy. Communications of the ACM, 61(8), 56–57. `https://doi.org/10.1145/3230718`.

Naylor, T. H. (1967). The guiding philosophy of an academic institution. Academe, 53(4), 305–309.

Nehme, A. (February 2023). The State of Data Literacy in 2023. Retrieved June 15, 2023, from `www.datacamp.com/blog/introducing-state-of-data-literacy-report`.

Newman, D. (May 21, 2018). Understanding The Six Pillars Of Digital Transformation Beyond Tech. Retrieved June 15, 2023, from `www.forbes.com/sites/danielnewman/2018/05/21/understanding-the-six-pillars-of-digital-transformation-beyond-tech/?sh=4776120c3f3b`.

Nimblett, M. (2018). Data literacy in higher education: The development of a data literacy program. Journal of Academic Librarianship, 44(6), 757–763. `https://doi.org/10.1016/j.acalib.2018.08.001`.

Okasha, A. (April 14, 2023). Overcoming Resistance: Strategies for Navigating Change in Digital Culture. Retrieved July 26, 2023, from `www.linkedin.com/pulse/overcoming-resistance-strategies-navigating-change-digital-okasha/`.

O'Neil, C. (2016). Weapons of Math Destruction: How Big Data Increases Inequality and Threatens Democracy. Broadway Books.

Overby, S. (April 26, 2023). How to Create a Data-Driven Culture. Wall Street Journal. Retrieved July 26, 2023, from `https://deloitte.wsj.com/articles/how-to-create-a-data-driven-culture-797c5e84?tesla=y&tesla=y`.

Pierce, J. G. (2010). Is the organizational culture of the U.S. Army congruent with the professional development of its senior level officer corps? Strategic Studies Institute, US Army War College. `www.jstor.org/stable/pdf/resrep12043.8.pdf`.

Pop, C. L. (2020). Fostering data culture through innovative digital storytelling. Information Services & Use, 40(3), 241–248. `https://doi.org/10.3233/ISU-200098`.

Quinn, R. E. and McGrath, M. R. (1985). The transformation of organizational cultures: A competing values perspective. In Frost, P. J., Moore, L. F., Louis, M. L., Lundberg, C. C., and Martin, J., eds., Organizational Culture, Sage Publications.

Rains, S. (October 27, 2022). Data Transformation and Change Management. BlueMargin. Accessed on July 26, 2023.

Reese, R. J. (2023). Enterprise Architecture in Action: Turning Business Strategy into Reality. ISBN: 9798854737074.

Rice, L. (March 11, 2023). How to Cultivate Communication in the Workplace. Qualtrics. Retrieved July 26, 2023, from www.qualtrics.com/blog/open-communication-culture/.

Robinson, A. (July 18, 2022) Collaboration: Definition, Examples & Tips. teambuilding.com. Retrieved from https://teambuilding.com/blog/collaboration.

Rouse, M. (May 8, 2017). What is a Data Governance Office (DGO)? – Definition from Techopedia. Retrieved August 1, 2023, from www.techopedia.com/definition/28038/data-governance-office-dgo.

Sainger, G. (2018). Leadership in Digital Age: A Study on the Role of Leader in this Era of Digital Transformation. International Journal on Leadership, 6(1), April. Retrieved on July 26, 2023.

SAS Institute Inc. (2018). Data literacy white paper. SAS Institute Inc.

Schein, E. (n.d). Quotes on Culture. Commisceo-global.com. www.commisceo-global.com/blog/what-is-your-favourite-quote-defining-culture.

Schlegelmilch, B. B. and Houston, M. J. (1989). Organizational changes: A study of the effects of a change in the organizational structure of a major U.S. retail chain. Journal of Marketing, 53(1), 15–32.

Sen, A. (April 29, 2019). Building a data-driven culture: The what, why, and how. DataCamp. Retrieved July 26, 2023, from www.datacamp.com/community/blog/building-a-data-driven-culture.

Shah, K. (June 4, 2020). Determining The Focus of Your Organization: The Identification Phase. Forbes.com. Retrieved from www.forbes.com/sites/forbescoachescouncil/2020/06/04/determining-the-focus-of-your-organization-the-identification-phase/.

Smith, R. (2009). Convergence or collision: The future of data and analytics. In Proceedings of the 25th Annual SAS Users Group International Conference (SUGI 25), Paper 064-25.

So, B. H.-H. and Swaminathan, S. (2018). The data-driven transformation: Opportunities and challenges. Business Horizons, 61(5), 697–706. https://doi.org/10.1016/j.bushor.2018.05.006.

Stanley, D. (2017). The data industry: The business and economics of information and big data. Routledge.

Sprouts. (2018). The design thinking process. YouTube. www.youtube.com/watch?v=_r0VX-aU_T8.

Stafford, J. (March 10, 2022). How to create the right culture to unlock value [Personal blog post]. Retrieved July 26, 2023, from www.ascent.io/blog/how-to-create-the-right-culture-to-unlock-data-value--/.

Statista (2023). Spending on digital transformation technologies and services worldwide from 2017 to 2026. *Statista.com.* www.statista.com/statistics/870924/worldwide-digital-transformation-market-size/.

Stevens, E. (July 26, 2021). 23 Must-read quotes about data. CareerFoundry. https://careerfoundry.com/en/blog/data-analytics/inspirational-data-quotes/.

Stitch (n.d.). The causes and costs of data silos. Stichdata.com. Retrieved from www.stitchdata.com/resources/data-silos/.

Statista (2023). Spending on digital transformation technologies and services worldwide from 2017 to 2026. Statista.com. www.statista.com/statistics/870924/worldwide-digital-transformation-market-size/.

Tan, A. (November 27, 2019). Data culture: What it is, why it's important, and how to build it. Tableau Software. Retrieved July 26, 2023, from www.tableau.com/learn/articles/building-data-culture.

The Strategy Group (2018). Design thinking in 90 seconds. YouTube. www.youtube.com/watch?v=vQytKCT563I.

Theodotou, M. (September 11, 2022). Leadership Blueprint: Cultivating Culture Change. eLearning Industry. Retrieved July 26, 2023, from https://elearningindustry.com/leadership-blueprint-cultivating-culture-change.

Tufte, E. R., (2001). The visual display of quantitative information. Cheshire, Conn.: Graphics Press.

Turriago-Hoyos, A., Thoene, U., and Arjoon, S. (2016). Knowledge Workers and Virtues in Peter Drucker's Management Theory. *SAGE Open, 6*(1). https://doi.org/10.1177/2158244016639631.

Tylor, Edward (1871). Primitive Culture. Vol 1. New York: J.P. Putnams Son; as referenced by Wikipedia (2023).

U.S. Department of Labor (n.d.). Office of Data Governance (ODG). Webpage on Office Overview. Retrieved August 1, 2023, from www.dol.gov/agencies/odg.

University of Minnesota (2015). Principles of Management. University of Minnesota Publishing. https://open.lib.umn.edu/principlesmanagement/chapter/8-5-creating-and-maintaining-organizational-culture-2/.

University of Minnesota (2017). Organizational Behavior. University of Minnesota Publishing. https://open.lib.umn.edu/organizationalbehavior/chapter/15-4-creating-and-maintaining-organizational-culture/.

Valladares, D., Sánchez, J. Á., and García, J. F. (2021). From Data to Knowledge: Fostering a Data Culture in Organizations. IEEE Software, 38(2), 92–99. https://doi.org/10.1109/MS.2020.3019649.

Vandenbosch, B., Geuens, M., and De Pelsmacker, P. (2014). The relationship between need for closure and cultural openness. International Journal of Research in Marketing, 31(4), 417–420.

Venema, M. (February 16, 2021). Espoused Values: Mind the Gap (and How to Close It). Harn.com. https://harkn.com/blog/espoused-values/.

Volpato, M. (March 24, 2022). How Data Leaders Can Overcome Resistance to Change [Podcast]. Retrieved July 26, 2023, from www.coriniumintelligence.com/content/podcasts/milton-volpato-digital-transformation-change-management.

Watkinson, A. and Kar, R. (March 24, 2023). Culture Transformation: What Leaders Need to Know. Gallup. Accessed on July 26, 2023 at www.gallup.com/workplace/471968/culture-transformation-leaders-need-know.aspx.

Wikipedia (2023). Culture. Wikipedia. https://en.wikipedia.org/wiki/Culture.

Wolff, A., Gooch, D., Cavero Montaner, J. J., Rashid, U., and Kortuem, G. (2016). Creating an understanding of data literacy for a data-driven society. The Journal of Community Informatics, 12(3), 9–26.

Zandvliet, J. (June 16, 2023). How to Create a Data Strategy that Succeeds. Blog Post. Retrieved from https://keyrus.com/us/en/insights/how-to-create-a-data-strategy-that-succeeds.

Zhu, X. and He, W. (2021). Building a data-driven culture for business growth. Business Horizons, 64(5), 729–739. https://doi.org/10.1016/j.bushor.2021.04.003.

Index

A

Artifacts, 97, 98, 103, 108, 189, 209

B

Business stakeholders, 154–156, 169

C

Change advisory board (CAB), 89,
    97, 98, 106
Chief Data Officer (CDO), 53, 54, 73–77,
    85, 88, 126, 149, 172
Coalition, 73, 74, 80, 81, 85
Communication, 98, 99, 116
    collaboration, 146, 189
        business stakeholders, 154–156
        data professionals, 154–156
        focus and collaboration, 146–148
        mature data, 148–152
        preservationist culture, 163–166
        process, 145
        progressive culture, 156–159
        protectionist culture, 165–168
        Silos, 152–154
        traditional culture, 160–163
        UFDCM, 156
    processes, 208
    systems, 196
Competing values model, 5
Culture, 57, 145
    and culture change, 7–10

data-driven, 54, 55
leadership, 73, 146
mature data (*see*
    Mature data culture)
organizational, 51
organizational decision-making, 49
perspective, 10–15
target data, 88
thrives, 189
transformation, 83
UFDCM (*see* Usage and Flow Data
    Culture Model (UFDCM))
    See also Data culture

D, E

Data accessibility, 63, 78, 79, 93, 117, 120,
    128, 129, 134, 154, 190, 209
Data analytics, 40, 46, 47, 70, 102, 103, 123,
    148, 151, 169
    capabilities, 54, 79
    data culture model, 125
    data-driven insights, 123
    data preservation, 124
    empowerment, 174
    internal decision-making, 63
    leveraging, 125–128
    preservationist culture, 124
    skills, 112
    tools, 176
    training, 83
    and visualization workshops, 58

© Gary W. Griffin, David Holcomb 2023
G. W. Griffin and D. Holcomb, *Building a Data Culture*, https://doi.org/10.1007/978-1-4842-9966-1

Data anonymization, 61, 70, 137
Data catalog, 93–95, 129, 149, 173
Data culture, 110, 111, 190, 203
    adoption, 180–182
    alignment, 2
    classified, 199
    continuum, 11
    in cultural transformation, 204
    DDDM, 179, 180
    decision-making, 172
    elements of, 3, 4
    embedding data, 123
    and governance, 88–91
    influences, 1
    initiative, 74
    interaction, 56–58
    KPI type, 175, 176
    LEA, 198
    leadership, 79
    leadership support, 172
    metrics and indicators, 171, 173–176
    organization, 74
    organizational and, 17
    organizational changes adaption, 184, 185
    organizational culture, 2
    organizational objectives, 176–178
    organizational structures (see Organizational structures)
    organization's performance, 1
    progress and success measuring, 178, 179
    strategies, 171
    strategy, 182, 183
    supportive environment, 83, 84
    survey, 60
    sustaining strategies, 185, 186
    transformation, 57–59, 82
    types, 171
    UFDCM (see Usage and Flow Data Culture Model (UFDCM))
Data culture maturity, 176–178, 199, 202, 207
Data culture model, 125, 130–132
Data-driven decision-making (DDDM), 11, 14, 19, 21, 49, 52, 53, 56, 58, 62, 71, 80, 118, 125, 151, 198
    and cross-functional environment, 78
    data integration, 172
    data utilization, 83
    evaluation, 179, 180
    KPIs, 175
    leadership, 73, 74
    ownership and leadership, 84
    pivotal role, 172
    transitions, 184
Data flow
    for effective analytics, 128, 129
    UFDCM, 130
Data governance, 10, 54, 58, 61, 64, 87, 125, 149
    communication, 98, 99
    compliance/security, 94–96
    and culture, 88–91
    data flow and usage, 101
    data quality, 197
    establishing, 63
    initiatives, 76, 197
    librarian initiatives, 91
    modern data catalog, 92–94
    organization, 77
    practices, 111
    preservationist culture, 102–104
    and privacy, 117
    processes, 70

process initiatives, 97, 98
program, 88, 201
progressive culture, 101, 102
protectionist culture, 105–107
quality, 96, 97
strategies, 76
structures, 155
traditional culture, 104, 105
training, 99, 100
UFDCM, 88
usage vector, 88
Data governance office (DGO), 38, 45, 97,
        148, 149, 154, 168
    establishment, 173
    mature data, 85
    organizational structures, 74–78
Data governance program (DGP), 38, 39,
        45, 149, 173
Data-informed decision-making
        (DIDM), 11, 56
Data inventory, 197
Data literacy, 40, 46, 100, 125, 148, 150,
        151, 172, 198, 199, 204
    continuous learning and data
        skills, 113–115
    culture, 109, 110
    data culture, 84
    and data-driven decision-making, 81
    data training programs, 111–113
    definition, 109
    importance, 73
    improvement, 171
    nurturing, 174
    in preservationist culture, 118, 119
    progressive culture, 119, 120
    protectionist culture, 116–118
    rate, 175
    and skills, 178

strategies, 85
strategies in data culture, 110–112
supporting, 66
in traditional culture, 115, 116
training programs, 58
and transparency, 78
triggers for, 120, 121
Data privacy
    measures, 131, 132
    and security, 130–132, 152
    usage, 132, 133
Data professionals, 154–156, 169
Data strategy, 37, 38, 45, 76, 88, 113, 148,
        168, 173, 204, 206
    communication, 54
    components, 51, 52
    and data culture interaction, 56–58
    data-driven decision-making, 53
    data flow, 53
    data-rich business
        environment, 52
    definition, 51
    highly regulated industries, 69–71
    implementation, 54
    organizational strategy, 53–55
    in preservationist culture, 61, 62
    progressive culture, 67–69
    protectionist culture, 62–64
    talent development, 59, 60
    traditional culture, 64–66
Data training programs, 67, 111–113, 119,
        171, 172, 178, 179
Data transformation, 51, 65, 79–81, 200
Data usage
    "additive" continuum, 60
    capabilities, 57
    and data flow, 42, 43
    data privacy and security, 132, 133

Data warehouse modernization, 201
Decline phase, 8
Democratization, 1, 37, 38, 41, 68, 98, 110, 133, 134
Digital transformation, 17, 199–201
Displacement, 12
Divisional organization structure, 28–30

F

Flat organizational structure, 33, 34
Flexibility, 5, 6, 15, 30, 98
Functional structures of organization, 26–28, 43, 78
Funding constraints, 201

G

Growth phase, 8, 42, 202, 203

H

Hierarchical organizational structure, 34, 35

I, J

Indicators, 173–176
Industries, 190
    case study 1, 197–199
    case study 2, 199–202
    case study 3, 202–206
    case study 4, 206–208
    preservationist culture, 193, 194
    progressive culture, 191, 192
    protectionist culture, 194–196
    traditional culture, 192, 193
Information and communications technology (ICT), 36, 88

Information technology (IT) organization, 77
Intellectual property (IP), 18, 152
Internal decision-making, 63–64
Inventory management, 127

K

Key performance indicators (KPI), 2
    characteristics, 174
    data culture type, 175, 176
    performance metrics, 180

L

Leadership, 74, 107, 113, 125, 126, 146, 153, 168, 169, 189, 190
    and advocacy, 172
    alignment, 197
    categories, 89
    change management, 82, 83
    coalition, 80, 81
    commitment, 185, 187
    data culture, 73
    data-driven decision-making, 74
    data privacy, 208
    data transformation, 79–81
    decision-making, 205
    designs, 17
    driving culture change, 79–81
    librarian services, 91
    organizational structure, 203
    policy and procedure, 73
    progressive culture, 77
    strategies, 74
    targeted data culture, 203
Local Education Agency (LEA), 197

M

Master data management (MDM), 10, 15, 93, 96, 98, 129, 204

Matrix organizational structure, 30, 31

Mature data culture, 51, 74, 78, 85, 88, 126, 171, 174, 202–204, 209

  clear and documented data processes, 39, 40

  communication, 148–152

  components assessment, 45–48

  data analytics, 40

  data literacy, 40

  data privacy and security, 41

  data strategy, 37, 38

  DGO, 38

  DGP, 38, 39

  governance, 96

Maturity phase, 8, 42

Metadata management, 10, 75, 89, 91, 92, 106, 129, 200, 201

Metrics, 173–176

Micro-cultures, 7, 25

Minimum viable product (MVP), 7, 206

Multi-factor authentication (MFA), 138

N

Need-to-know culture, 12, 13, 18, 24, 27, 106

Network organization structure, 31, 32

O

Operational data store (ODS), 200

Organizational culture, 4–7, 42, 71, 130, 190, 202, 205, 209

Organizational life cycle, 7, 8, 36, 42, 43, 120, 202

Organizational strategy, 54, 55

Organizational structures, 91, 205

  assessment, 43–45

  and DGO, 74–78

  divisional, 28–30

  flat, 33, 34

  functional, 26–28

  hierarchical, 34, 35

  horizontal and vertical, 26

  matrix, 30, 31

  network, 31, 32

  team-based, 32, 33

  UFDCM, 26

  virtual work, 35, 36

P, Q

Personally identifiable information (PII), 88, 95, 130

Preservationist data culture, 26, 78, 79, 81, 84

  analytics, privacy, and security, 138–140

  communication and collaboration, 163–166

  data governance, 102–104

  data strategy, 61, 62

  healthcare and finance, 209

  industries, 193, 194

  literacy, 118, 119

  UFDCM, 25, 26

Privacy-preserving techniques, 137

Progressive data culture, 79, 81, 84

  analytics, privacy, and security, 140–142

  communication and collaboration, 156–159

Progressive data culture (*cont.*)
    data governance, 101, 102
    data strategy, 67–69
    establishment, 201
    evolution and change, 209
    industries, 191, 192
    literacy, 119, 120
    UFDCM, 19–24
Protectionist data culture, 78, 79, 81, 84
    analytics, privacy, and
        security, 136–138
    communication and
        collaboration, 165–168
    data governance, 105–107
    data strategy, 62–64
    industries, 194–196
    literacy, 116–118
    UFDCM, 24

R

Remote work, 35
Resistance, 81, 82, 189

S

Self-managed teams, 32, 33
Silos, 152–154
Skepticism, 81, 82, 110
Startups, 5, 7, 191, 206
Strategic Road Map, 48, 49, 173
Subcultures, 7

T

Talent development, 52, 59, 60, 71
Team-based structure, 32, 33, 48
Telework, 35

Trade-offs, 6, 26
Traditional culture, 21–24, 56, 64–66
Traditional data culture, 79, 81, 104, 105
    analytics, privacy, and
        security, 134–136
    communication and
        collaboration, 160–163
    data literacy, 115, 116
    industries, 192, 193
Transformation, 57–59, 171, 172, 181
Transition, 9, 197, 205
Trust, 11, 110, 133, 134, 143

U

Usage and Flow Data Culture Model
    (UFDCM), 2, 13, 15
    alignment, 199
    classification, 201
    collaboration and
        communication, 145
    culture type, 18
    data culture, 199
    data culture types, 56, 113, 171
    data flow vector, 152
    data strategy, 51
    data usage and flow, 77
    definition, 18
    industries, 190
    leadership, 80, 153
    organization's culture, 18
    preservationist culture, 25, 26, 61, 62
    progressive culture, 19–21
    protectionist culture, 24
    quadrants, 52
    security, 133
    traditional culture, 21–24
    valuable perspective, 42

V

Vectors, 5, 6, 57, 60
    data accessibility, 39
    data flow, 22, 24, 37, 39, 75, 124
    data-handling practices, 130
    data usage, 22, 23, 37, 125
    self-imposed limitations, 166
    UFDCM, 38
Virtual work, 35, 36

W, X, Y, Z

Work from home, 35